Seismic Evaluation and Design of Petrochemical and Other Industrial Facilities

Third Edition

Prepared by
Task Committee on Seismic Evaluation and Design of
Petrochemical Facilities of the
Energy Division of the
American Society of Civil Engineers

Published by the American Society of Civil Engineers

Library of Congress Cataloging-in-Publication Data

Names: American Society of Civil Engineers. Task Committee on Seismic Evaluation and Design of Petrochemical Facilities, author.
Title: Seismic evaluation and design of petrochemical and other industrial facilities / prepared by Task Committee on Seismic Evaluation and Design of Petrochemical Facilities of the Energy Division of the American Society of Civil Engineers.
Other titles: Guidelines for seismic evaluation and design of petrochemical facilities
Description: Third edition. | Reston : American Society of Civil Engineers, [2020] | Revision of Guidelines for seismic evaluation and design of petrochemical facilities. | Includes bibliographical references and index. | Summary: "This report offers practical recommendations regarding the design and safety of new and existing petrochemical facilities during and following an earthquake"– Provided by publisher.
Identifiers: LCCN 2019041768 | ISBN 9780784415481 (print) | ISBN 9780784482667 (PDF)
Subjects: LCSH: Petroleum refineries–Design and construction. | Earthquake resistant design–Standards–United States.
Classification: LCC TH4571 .G85 2020 | DDC 665.5/3–dc23
LC record available at https://lccn.loc.gov/2019041768

Published by American Society of Civil Engineers
1801 Alexander Bell Drive
Reston, Virginia 20191-4382
www.asce.org/bookstore | ascelibrary.org

ASCE Petrochemical Energy Committee

This publication is one of five state-of-the-practice engineering reports produced to date by the ASCE Petrochemical Energy Committee. These engineering reports are intended to summarize current engineering knowledge and design practice and present guidelines for the design of petrochemical facilities. They represent a consensus opinion of the task committee members who are active in their development. These five ASCE engineering reports are

1. *Design of Anchor Bolts in Petrochemical Facilities,*
2. *Design of Blast Resistant Buildings in Petrochemical Facilities,*
3. *Design of Secondary Containment in Petrochemical Facilities,*
4. *Seismic Evaluation and Design of Petrochemical and Other Industrial Facilities*
   (Note:First and second editions were titled *Guidelines for Seismic Evaluation and Design of Petrochemical Facilities,* and the name was modified for the third edition at the request of ASCE Publications), and
5. *Wind Loads for Petrochemical and Other Industrial Facilities.*

A. K. Gupta organized the ASCE Petrochemical Energy Committee in 1991, which was initially chaired by Curley Turner. Under their leadership the five task committees were formed, initially publishing the five reports in 1997. The Committee was subsequently chaired by Joseph A. Bohinsky and Frank J. Hsiu. In 2005, Magdy H. Hanna reorganized the ASCE Petrochemical Energy Committee, and the following four task committees were formed to update their respective reports:

- Task Committee on Anchorage Design,
- Task Committee on Blast-Resistant Design,
- Task Committee on Seismic Evaluation and Design for Petrochemical Facilities, and
- Task Committee for Wind-Induced Forces.

Building codes and standards have changed significantly since the publication of these five reports, specifically in the calculation of wind and seismic loads and analysis procedures for anchorage design. In addition, new research in these areas and in blast-resistant design has provided opportunities to improve the recommended guidelines. ASCE has determined the need to update two of the original reports and publish new editions based on the latest research and for consistency with current building codes and standards.

In 2014, the Energy Division Executive Committee Chair J. G. (Greg) Soules requested the following two task committees to update their respective reports:

- Task Committee on Seismic Evaluation and Design of Petrochemical Facilities, and
- Task Committee for Wind-Induced Forces.

Current ASCE Petrochemical Energy Committee

James R. (Bob) Bailey   Exponent—Chairman
J. G. (Greg) Soules   CB&I Storage Tank Solutions LLC

The ASCE Task Committee on Seismic Evaluation and Design of Petrochemical Facilities

This revised document was prepared to provide guidance in the seismic design of new petrochemical and other industrial facilities and the seismic evaluation of existing facilities.Though the committee membership and the intent of this document are directed to petrochemical facilities, these guidelines are applicable to similar situations in other industries. The intended audience for this document includes structural design engineers, operating company personnel responsible for establishing seismic design and construction standards, and local building authorities.

The task committee was established because of the petrochemical industry's significant interest in addressing the wide variation of petrochemical-industry-related design and construction practices and standards that are applied throughout the country.Another primary purpose was to address the need for consistent evaluation methodologies and standards for existing facilities. Most governing building codes and design standards address only new design, and clearly retro-fitting existing facilities to meet current standards would be prohibitively expensive. Furthermore, standards for new design do not address all of the conditions that may be found in existing facilities.

These guidelines are intended to provide practical recommendations in several areas that affect the safety of a petrochemical facility during and following an earthquake.

In the area of new design, these guidelines emphasize interpretations of the intent of building codes as applied to petrochemical facilities and practical guidance on design details and considerations that are not included in building codes.

For existing facilities, these guidelines provide evaluation methodologies that rely heavily on experience from past earthquakes, coupled with focused analyses. The guidelines emphasize methods to address seismic vulnerabilities that building codes do not cover, but that experienced engineers can identify.

This book also provides background information and recommendations in several areas related to seismic safety where the structural engineer may interact with other disciplines and with plant operations. These areas include seismic hazards, contingency planning, and post-earthquake damage assessment.

The original version of this document, published in 1997, was developed by a committee of industry representatives chaired by Mr. Gayle S. Johnson. A reconstituted committee led by Mr. J. G. (Greg) Soules created the second edition, published in 2011.

For this third edition, several key individuals dedicated significant amounts of time to formulating, writing, and reviewing in detail specific sections of this document. Those members are identified as follows.

The ASCE Task Committee on Seismic Evaluation and Design of Petrochemical Facilities

J. G. (Greg) Soules, P.E., S.E., SECB
CB&I Storage Tank Solutions LLC
Chair

Gayle S. Johnson, P.E.   Simpson Gumpertz & Heger Inc.
Ahmed Nisar, P.E.   InfraTerra, Inc.

The committee would like to thank the following individuals for their reviews and other contributions.

Reviewers and Other Contributors
Robert E. Bachman, S.E.   Robert Bachman Consulting
Michael W. Greenfield, Ph.D., P.E.   Greenfield Geotechnical LLC
Christopher Hitchcock, CEG   InfraTerra, Inc.
Justin D. Reynolds, P.E.   Simpson Gumpertz & Heger Inc.
Paul B. Summers, S.E.   Simpson Gumpertz & Heger Inc.
Guzhao Li, Ph.D., S.E.   Simpson Gumpertz & Heger Inc.

Contents

CHAPTER 1

Introduction

1.1 OBJECTIVE

Many codes and standards are used in the structural and seismic design and assessment of petrochemical facilities. Many of these codes were developed primarily for use in designing buildings and generally offer insufficiently detailed guidelines for complete design and evaluation of structures commonly found in petrochemical and other industrial facilities. As a result, the engineer is often forced to rely on broad subjective interpretation of these codes' intent to develop detailed design criteria and procedures that apply to items found in petrochemical facilities. Many petrochemical operating companies with facilities in seismic regions, and engineering offices that serve the petrochemical industry, have developed their own internal standards and guidelines for addressing these unique seismic design and evaluation issues. Consequently, these facilities may be designed and built with inconsistent degrees of conservatism and design margins.

Until the publication of the first edition of this book, no widely accepted standards for the seismic evaluation of existing facilities existed. As the public and regulators become more aware of environmental and safety issues associated with such facilities, the need for regulators, owners, and engineers to have a consistent approach and a technically sound, practical basis for performing evaluations grows continually.

Recognizing the need for design and evaluation guidelines in several technical areas specifically applicable to petrochemical facilities, ASCE's Energy Division set up the Petrochemical Energy Committee to fill the existing gap by establishing criteria and guidance for practical application. The Task Committee on Seismic Evaluation and Design of Petrochemical Facilities of the Petrochemical Energy Committee was charged with development of the original document, which was first published in 1997. The original document was based on the seismic requirements of the 1994 *Uniform Building Code* (UBC 1994). Historically, the UBC had used the Structural Engineers Association of California's "Recommended Lateral Force Requirements and Commentary" (SEAOC 1999), also known as the "Blue Book," as the basis for its seismic provisions. The second edition of this book was developed by a reconstituted Seismic Task Committee and was based on the 2006 *International Building Code* (IBC 2006), which adopted, by

1

reference, the seismic provisions of ASCE 7-05 *Minimum Design Loads and Associated Criteria for Buildings and Other Structures, including Supplement 2* (ASCE 2005). This third edition was revised by another reconstituted Seismic Task Committee and is based on the 2018 *International Building Code* (IBC 2018), which adopts, by reference, the seismic provisions of ASCE Standard 7-16 (ASCE 2016). Whereas the 2018 IBC does not reference Supplement 1 to ASCE 7-16, the corrected site coefficients in Supplement 1 are critical for the proper seismic design of structures; thus, this book refers to Supplement 1.

This document has been developed to provide practical guidance to engineers involved in the seismic design and evaluation of petrochemical and other industrial facilities. It aims to serve several objectives:

(a) To help practicing engineers better understand the intent of certain provisions of seismic design codes, enabling them to apply the codes and provisions more properly and uniformly to structures and systems typically found in petrochemical and other industrial facilities;

(b) To provide guidance for seismic engineering practice beyond that covered in the building codes;

(c) To provide background information on technical areas that are related to the seismic evaluation of petrochemical and other industrial facilities, but that civil engineers do not always understand well;

(d) To provide guidance specific to the seismic evaluation of existing petrochemical and other industrial facilities;

(e) To provide practical analytical guidance specifically applicable to petrochemical and other industrial facilities; and

(f) To alert engineers and operations personnel to areas other than structural where earthquakes might affect the safety of petrochemical and other industrial facilities, such as contingency planning, post-earthquake damage inspection, and operational issues.

This book addresses seismic design and related construction of new structures and components as well as evaluation and retrofit design of existing structures and systems in petrochemical facilities. The scope generally emphasizes work that is commonly under the direction of an engineer.

Despite the attempt to make this a comprehensive document, the authors recognize that some applicable topics will not be fully covered.

Certain types of petrochemical facilities must adhere to specific government requirements. For example, liquified natural gas (LNG) terminals in the United States are regulated by the Federal Energy Regulatory Commission (FERC). FERC generally requires conformance to specific documents such as NFPA 59A (NFPA 2001, 2016), API 620 (API 2014), and ASCE 7 for seismic design. Much of the guidance in this book, such as the discussion on site-specific ground motion in Chapter 3, is directly applicable, although FERC may impose additional requirements.

Note that this book is not intended to replace the IBC and ASCE 7 but is intended to be used in conjunction with both. In all instances, the engineers involved in the design or evaluation process are ultimately responsible for researching the literature and using their professional judgment to ensure that applicable safety objectives, criteria, and other performance goals are met.

1.2 RELATED INDUSTRY CODES, STANDARDS, AND SPECIFICATIONS

Several industry codes, standards, and specifications are normally used for the design of structures and components at petrochemical facilities. Applicable codes, standards, and specifications may include, but are not limited to the following:

(a) 2018 IBC, *International Building Code*;

(b) ASCE/SEI 7-16, *Minimum Design Loads and Associated Criteria for Buildings and Other Structures*, Including Supplement No. 1;

(c) ANSI/AISC 341-16, *Seismic Provisions for Structural Steel Buildings*;

(d) ANSI/AISC 360-16, *Specification for Structural Steel Buildings*;

(e) ACI 318-2014, *Building Code Requirements for Structural Concrete*;

(f) TMS 402-2016, *Building Code Requirements for Masonry Structures*;

(g) TMS 602-2016, *Specification for Masonry Structures*;

(h) API 650-2014, *Welded Steel Tanks for Oil Storage*, 12th Edition, Addendum 1;

(i) ASME B31.3-2014, *Process Piping*;

(j) AWC NDS-2018, *National Design Standard for Wood Construction*; and

(k) AWC SDPWS-2015, *Special Design Provisions for Wind and Seismic*.

With respect to code provisions, the guidelines presented in this book often refer to the 2018 IBC, because the IBC, by reference, is the most widely used design code in the United States. Whereas the IBC requires the use of ASCE 7 to determine seismic forces and their distribution in a structure, the load combinations of the 2018 IBC include alternative basic load combinations for allowable stress design that may prove more favorable to the user for stability sizing of foundations for industrial equipment. The basic load combinations of both the IBC and ASCE 7 may produce foundation sizes that are significantly larger than those historically designed in petrochemical and other industrial facilities.

Note that ongoing efforts in code development and refinement related to the IBC and other codes may at one time or another make some of the specific guidance provided in this book obsolete, especially where related to particular building code provisions. However, the intent and philosophy of much of the guidance provided herein is expected, for the most part, to remain directly applicable to petrochemical and other industrial applications. Much of the design philosophy, analytical techniques, analytical tools, and specific guidance provided in this book are appropriate across a wide range of design criteria.

Note also that the guidelines provided herein are not intended to take precedence over the code of record wherever differences may occur. They are, however, intended to provide a rational basis for deviating from certain provisions of standards in establishing specific design requirements for items particular to petrochemical facilities. Where differences exist between the guidelines of this book and the code of record, the latter should always prevail, unless a potential deviation from the standard code of practice is approved by the Authority Having Jurisdiction.

1.3 ORGANIZATION OF THE BOOK

The remainder of the book has been organized into chapters that address specific seismic-related aspects of petrochemical facility design, evaluation, or operations. The following paragraphs briefly summarize the contents of each chapter.

Chapter 2 discusses seismic design philosophy and the general intent of seismic design provisions. Performance requirements are also discussed.

Chapter 3 provides background data related to seismic hazards, such as ground shaking, fault rupture, and tsunamis, and geotechnical issues, such as liquefaction. The chapter emphasizes information that will help the engineer understand the derivation and significance of different definitions of ground motion that may be encountered on a project.

Chapter 4 addresses analysis and load definition for petrochemical facilities, providing guidance on interpreting building code provisions for application to the types of structures commonly found in petrochemical facilities. The appendixes to Chapter 4 offer practical analytical tools for several necessary tasks in facility structural design and evaluations, such as calculation of structural periods for components unique to petrochemical facilities, calculation of sliding displacements, and stability checks.

Chapter 5 provides guidance for the design of new components in petrochemical and other industrial facilities. This chapter offers useful guidance from experienced engineers, emphasizing interpretation of the intent of code provisions and specific items not found in typical design codes and standards, such as special design details and configuration controls.

Chapter 6 relates specifically to the evaluation of existing facilities, providing detailed guidance on methodical "walkdown" screening reviews of structures and systems.

Chapter 7 covers the evaluation of flat-bottomed steel storage tanks. The chapter discusses different design codes and presents alternative methodologies that are particularly useful for the evaluation of existing tanks.

Chapter 8 addresses earthquake contingency planning for a facility. It does not give specific guidance on how to author contingency plans; rather, it outlines some general points that in-place contingency plans should address.

Chapter 9 provides guidance on post-earthquake damage assessment of petrochemical facilities.

Chapter 10 provides guidance to the engineer faced with retrofitting seismically deficient structures in a petrochemical facility.

Chapter 11 discusses topics involved in the seismic design of marine oil terminals.

Chapter 12 provides guidance in evaluating multiple codes on a project, in particular as applied to international projects and vendors.

References

ACI (American Concrete Institute). 2014. *Building code requirements for structural concrete.* ACI 318. Farmington Hills, MI: ACI.

ACI. 2016a. *Building code requirements for masonry structures.* TMS 402. Farmington Hills, MI: ACI.

ACI. 2016b. *Specifications for masonry structures.* TMS 602. Farmington Hills, MI: ACI.

AISC. 2016a. *Seismic provisions for structural steel buildings.* AISC 341. Chicago: AISC.

AISC. 2016b. *Specification for structural steel buildings.* AISC 360. Chicago: AISC.

API (American Petroleum Institute). 2014a. *Design and construction of large, welded, low pressure storage tanks,* 12th ed. Washington, DC: API.

API. 2014b. *Welded steel tanks for oil storage,* 12th ed. Washington, DC: API.

ASCE. 2005. *Minimum design loads for buildings and other structures, including supplement 2.* ASCE 7-05. Reston, VA: ASCE.

ASCE. 2010. *Minimum design loads for buildings and other structures, including supplement 2.* ASCE 7-10. Reston, VA: ASCE.

ASCE. 2016. *Minimum design loads and associated criteria for buildings and other structures.* ASCE/SEI 7-16. Reston, VA: ASCE.

ASME. 2014. *Process piping.* ASME B31.3. New York: ASME.

AWC (American Wood Council). 2015. *Special design provisions for wind and seismic.* AWC SDPWS-15. Leesburg, VA: AWC.

AWC. 2018. *National design standard for wood construction.* AWC NDS-18. Leesburg, VA: AWC.

IBC (International Building Code). 2006. *International building code.* Country Club Hills, IL: IBC.

IBC. 2018. *International building code.* Country Club Hills, IL: IBC.

NFPA (National Fire Protection Association). 2001. *Production, storage, and handling of liquefied natural gas (LNG).* NFPA 59A. Quincy, MA: NFPA.

NFPA. 2016. *Production, storage, and handling of liquefied natural gas (LNG).* NFPA 59A. Quincy, MA: NFPA.

SEAOC. 1999. *Recommended lateral force requirements and commentary.* Sacramento, CA: SEAOC.

UBC (Uniform building code). 1994. *Uniform building code.* Whittier, CA: International Conference of Building Officials.

CHAPTER 2

Design and Evaluation Philosophy

2.1 INTRODUCTION

This chapter presents the broad design philosophy underlying this document. The design philosophy derives from current knowledge of seismic hazards, observed and recorded behavior of structures and components during earthquakes, and consideration of consequences of failure of these structures and components. Available documents on recommended seismic design practices for conventional and special facilities (ASCE 2016, FEMA 2015) were used in the overall development of design and evaluation philosophies.

The overall philosophy of this guidance document is based on the premise that proper seismic design can be achieved by

  (a) Determining all applicable seismic hazards for a site,

  (b) Defining performance objectives for different usage categories of structures and components,

  (c) Establishing design bases that meet the prescribed seismic performance objectives for these facilities, and

  (d) Ensuring that the design and construction adhere to those bases.

2.2 CONSIDERATIONS FOR NEW DESIGN

The design philosophy presented for new facilities assumes that the code of record is the 2018 IBC. Use of these guidelines with other similar codes and standards that may be in effect in a particular community should result in comparable designs.

The seismic design forces that this document discusses are based on the assumption that ductility is provided in new design and that inelastic behavior will occur in structures and components during strong ground motions. As a result, the lateral design forces are significantly smaller than those that would be required if the structures and components were designed to remain elastic.

For structures in the inelastic regime, structural behavior improves if the inelastic deformations are well distributed throughout the structure. The seismic force levels called for in this document are based on this consideration. Higher elastic force reduction factors are thus provided for cases where a more uniform distribution of inelastic deformation can be developed in the lateral force–resisting system (e.g., a well-distributed system of moment-resisting frames). Conversely, for cases where inelastic deformation may concentrate in a few members, the reduction factors are lower (e.g., elevated tanks). In any case, the engineer is responsible for developing a design that will correspond to the expected inelastic response capability of the structural system for the item under consideration.

Furthermore, the guidance provided in this document generally provides minimum design criteria. The engineer must interpret and adapt these guidelines to each item using experience and judgment. In selected cases, adopting more conservative criteria that may have overall long-range economic benefits may be advantageous. Because of the great variability and options available in the design process and because of the complexity of structures and components, covering all possible variations in seismic response and providing optional detailed criteria is very difficult. Thus, the engineer has both latitude and responsibility to exercise judgment in the development of detailed design criteria and in the execution of the design.

2.3 CONSIDERATIONS FOR THE EVALUATION OF EXISTING FACILITIES

The evaluation of existing facilities may involve a different design basis or different analysis techniques and acceptance criteria than used for new design, depending on the regulatory requirements, or, in case of a voluntary upgrade, the performance criteria established by the owner. Several key considerations are as follows:

(a) Evaluation acceptance criteria (stress and/or deformation limits) may be more or less conservative than those for an equivalent new facility depending on the intent of the evaluation.

(b) Additional requirements related to equipment functionality and systems interaction may be appropriate.

(c) Loads on an existing structure may have changed over time, and the assessment should consider the actual loads.

(d) Any changes in the operating basis (weight, operating conditions, etc.) may affect the assessment and should be incorporated.

(e) Remaining design life.

(f) In evaluating existing facilities, actual material properties (if available) should be taken into account.

(g) Information from available field observations, such as deterioration of the structural elements (e.g., corrosion) or missing or changed structural elements, should be addressed.

In summary, the "best estimate" of the structure and material properties should be used to get the most accurate prediction of the structure's or system's performance. Anything that might affect system performance should be considered, whether structural or functional.

2.4 CAUTIONS REGARDING DESIGN AND EVALUATION

The current building code philosophy is to design a building or structure for life safety when subjected to the design seismic event. The building or structure is designed to maintain its structural integrity and stability during the design seismic event, but significant structural damage will likely occur during a major earthquake. Clear communication is needed to ensure that the owners and regulators understand that structural damage is expected and that repairs will likely be required after a design event.

Seismic risk can never be completely eliminated. Although conservatism can be added to the design and/or review criteria, and a more detailed and extensive investigation (at greater cost) may yield more accuracy and reliable information, some level of seismic risk will always exist. This is true regardless of how much time and resources are spent in the design or retrofit of a facility. For the evaluation of existing structures and for voluntary upgrades, the goal should always be to "minimize the risk" given available resources. To avoid conflicting expectations among engineers, owners, regulators, or other affected parties, the following points should be discussed openly and possibly agreed to in writing to avoid misunderstandings at later dates:

(a) All parties must recognize the lack of complete assurance of meeting the desired performance criteria during a design event, particularly when seismic evaluations are conducted for existing facilities. Besides the uncertainties associated with material properties and structural behavior (uncertainties that are typically larger for existing facilities than for new designs), large uncertainties are associated with the earthquake input motion in terms of amplitude, frequency content, and duration.

(b) The engineer has an obligation to use the degree of care and skill ordinarily exercised by reputable, experienced engineering professionals in the same or similar locality and under similar circumstances at the time the evaluation is performed

(c) All parties must recognize that geologic, seismic, environmental, structural, and geotechnical conditions can vary from those encountered when and where the engineer obtained the initial data used in the evaluation and that the limited nature of the data necessarily causes some level of uncertainty with respect to its interpretation, notwithstanding the exercise of due professional care.

(d) All parties must recognize that the extent of the engineer's evaluations is always limited by the time frame and funds available for the investigation. A more detailed and extensive investigation, at greater cost, might yield more accurate and reliable information that might affect some of the engineer's decisions and judgments.

(e) When evaluating existing facilities, all parties must agree on reasonable limits to the engineers' design responsibility and liability with regard to the adequacy of the original design and construction. Normally, the evaluating engineer should not assume responsibility for the original design and construction. Owners and engineers should be made aware of extensive uncertainties that may remain in the absence of significant investigations.

This document is also intended to reflect current common practice and should not limit the engineer from developing designs and retrofits consistent with emerging knowledge.

2.5 PERFORMANCE OBJECTIVES AND RISK CATEGORIES

2.5.1 Performance Objectives

The criteria in this document are based on performance objectives defined for selected risk categories. The performance objectives are intended to provide different levels of protection from damage, consistent with the risk categories. The main elements of these performance objectives are as follows:

(a) Structural integrity: The design philosophy articulated in this document calls for maintaining the structural integrity of all structures and components in a facility. This means that structures and components should not collapse or otherwise fail under the design basis ground motion specified for the site. Functionality may need to be maintained if specifically required by the owner or jurisdictional authorities or if required to protect public health and safety. (In general, maintaining structural integrity does not imply maintaining functionality.) Maintaining structural integrity requires attention to strength, ductility, and deformation limits. Subsequent chapters provide guidance for strength, ductility, and deformation limits. Although the guidance presented in this document is not intended to prevent damage, much of the guidance is intended to limit damage that would adversely affect public safety.

(b) Containment: Structures, systems, and components with hazardous materials should be designed to contain such materials during and after a major earthquake. Ensuring containment requires attention to strength, ductility, and deformation limits and to structural details of elements with respect to potential leak paths. Preventing the release of significant quantities of hazardous materials into the environment is essential to avoid endangering facility personnel and to maintain public health and safety in the event of a major earthquake.

(c) Functionality: All structures, systems, and components that are needed after an earthquake, such as fire prevention or other emergency systems, should be designed to maintain functionality (i.e., continued operability during or after the design earthquake). Maintaining functionality requires attention to strength, ductility, and deformation limits and to stress levels, structural details, seismic interaction, and protection of essential systems and components.

2.5.2 Risk Categories

For purposes of seismic design and/or evaluation, structures and components at a petrochemical or other industrial facility are divided into four risk categories: low hazard to human life (I), substantial hazard to human life (III), essential facilities (IV), and other structures (II).

Category I covers buildings or other structures with low hazard to human life in the event of failure. This includes barns, storage shelters, gatehouses, and similar small structures.

Category III includes buildings and structures that represent a substantial hazard to human life in the event of a failure. This includes specific high-occupancy buildings, for example, schools with capacities greater than a defined limit, and structures containing sufficient quantities of hazardous materials that, if these materials were released, would be dangerous to the public. Structures and components in this category should be designed to maintain their integrity and to contain the hazardous material. They need not maintain their general functionality beyond what is required for integrity or containment. If these structures and components fail, they should fail in a manner that will preclude release of hazardous material into the environment.

Category IV includes essential facilities, such as fire stations and designated emergency response centers, and buildings and other structures containing extremely hazardous materials where the quantity contained exceeds a threshold quantity established by the Authority Having Jurisdiction. Structures and components in this category are designed to maintain their functionality after an earthquake.

Category II includes all buildings and structures that are not included in Categories I, III, and IV.

2.5.3 Relationships between Performance Objectives and Risk Categories

Table 2-1 shows the relationships between performance objectives and risk categories. Design for all risk categories should provide for structural integrity. In addition, hazardous material–handling facilities and components should be designed to ensure containment of such hazardous material. Similarly, essential facilities and certain structures and components in the hazardous material–handling facilities should be designed to maintain selected functionality. Finally, the qualitative seismic risk for each risk category should be consistent with the philosophy discussed in this section, as shown in the last column of Table 2-1. A petrochemical or other industrial facility will likely have structures and buildings that fall within different risk categories. Critical firefighting and central process control facilities are examples of structures that fall within the essential risk category (Category IV). Process units handling inert solids may be classified as Category II. The owner or the local building official should review the risk category to be used for specific units, structures, or buildings.

2.5.4 Basis for the Recommended Performance Objectives

The underlying basis for any seismic design criterion or procedure is the specification, either explicit or implicit, of an acceptable level of risk. This document was prepared with the belief that, although not quantified, the performance objectives specified for petrochemical and other industrial facilities should be consistent with public safety expectations of other civil structures. In other words, the risk of not meeting the performance objectives should not exceed commonly accepted risks associated with major commercial/industrial facilities whose failure may affect public health and safety.

Table 2-1. Performance Objectives for Different Risk Categories.

	Performance Objectives			
Risk Categories	Maintain Structural Integrity	Contain Hazardous Materials	Maintain Functions	Qualitative Seismic Risk
I Low Hazard	Yes	N/A	No	Low
II Other	Yes	No[4]	No[1]	Low
III Substantial Hazard	Yes	Yes	No[2]	Very Low
IV Essential	Yes	Yes	Yes[3]	Extremely Low

[1]Owners may choose items designed to maintain function with a significant economic impact.
[2]Maintain only those functions that are required for containment.
[3]Maintain only those functions that are required for essential operations or for containment.
[4]This category would be used with hazardous materials provided that the requirements of ASCE 7-16, Table 1.5-1 and Section 1.5.3 are met.

Furthermore, seismic risk should be consistent with the potential consequences of failure of structures and components; those items that are likely to pose greater hazard to life safety should have a lower risk of not meeting their performance objectives. This risk is based on current knowledge of seismic behavior and is not quantified.

2.6 DESIGN APPROACH FOR NEW FACILITIES

Engineering requirements and acceptance criteria that demonstrate compliance of each risk category with applicable performance requirements should be based on well-established approaches that are consistent with current seismic design practices. The major elements of these criteria include

(a) Strength: The structural design should be such that risk of gross failure of the structural systems and elements under the design loads is minimized. This is accomplished by demonstrating that the strength, based on the acceptance criteria, is equal to or greater than the demand resulting from the design loads. In determining the design loads, the loading combinations should account for all likely conditions, consistent with current practices.

(b) Ductility: Ductility may be relied upon in determining the capability of structures and components to resist seismic loads. The ductility limits and the accompanying deformations should not exceed the limiting values found herein, consistent with the previously described risk categories. Ductility limits may also be demonstrated through appropriate independent testing.

(c) Seismic analysis and design methods: As discussed in detail in Chapter 4, the equivalent lateral force (ELF) method of analysis is generally appropriate for seismic analysis and design. However, specific systems may require dynamic analysis to better understand seismic structural response, including a more accurate distribution of seismic forces. The engineer should use judgment in deciding whether a dynamic analysis is needed, except in cases where the codes explicitly require such dynamic analysis (e.g., irregular structures). Consideration should be given to the fact that the formulations in the ELF method are generally empirical and are based primarily on building structures. At petrochemical and other industrial facilities, the structures in many cases are different even if they qualify as "regular" and they may involve interaction with piping systems of significant mass. In such cases, a dynamic analysis may be a better choice.

Both ELF and dynamic analyses should be performed using linear elastic methods. Seismic loads should be properly combined with the other load effects to obtain the "design forces" in accordance with the project criteria.

(d) Acceptance criteria: Either allowable stress or ultimate strength design methods may be used in establishing the acceptance criteria. In addition, where appropriate, displacement limitations should be imposed to ensure acceptable behavior of structures and components.

(e) Seismic review of systems and components: Experience indicates that many systems and types of equipment are generally seismically rugged, that is, they are ordinarily constructed such that they have inherent seismic capacity to survive strong motions without significant loss of function. The engineer should assess the systems and components, especially those in the hazardous and essential categories, to determine their ruggedness. Those systems and components that are deemed to be susceptible to strong ground shaking should be reviewed in more detail (see ASCE 7-16, Chapter 13 for guidance on equipment qualification). Such a review may be demonstrated by experience data, analysis, testing, or any combination thereof.

(f) Redundancy: Experience indicates that redundancy is a significant factor affecting the survivability of a structure or component during strong ground motion. Many structures and components at petrochemical and other industrial facilities lack redundancy. Consideration should be given to providing redundant structural systems to the maximum extent practicable. Although providing redundancy may result in a small increase in the construction costs, it will also reduce the seismic risk by providing alternate load paths and additional energy dissipation mechanisms.

(g) Seismic interaction: Seismic interaction, such as impact or differential displacement, may have adverse effects on the performance of sensitive systems and components. Therefore, separation of such items to preclude seismic interaction with other nearby structures and components is a basic design principle. If separation is not practical, then seismic interaction should be accounted for in design through analysis.

The guidance provided in this document is generally based on the elements mentioned above. Implementing these design principles should result in a design that will meet the performance objectives. In particular, both strength and ductility requirements should be specified to meet the structural integrity, containment, or functionality criteria. As expected, the permissible ductility values listed in later sections of this document are relatively high and apply to the performance criterion of maintaining structural integrity. They would necessarily be lower if a performance criterion of functionality were specified.

2.7 EVALUATION OF EXISTING STRUCTURE/REPLACEMENT IN KIND

Chapter 10 provides guidance for evaluating and retrofitting a seismically deficient structure, providing examples of situations that would trigger a need to evaluate and potentially upgrade an existing structure.

This section addresses the situation in which an existing piece of equipment (e.g., drum, tower, or exchanger) is replaced by a new, identical piece of equipment—that is, a replacement in kind. A replacement in kind, by definition, involves neither a change in usage, nor an addition to or modification of an existing structure.

For a replacement in kind, the equipment supplier will design the new equipment per the current design codes and will provide new structure or foundation loads (wind, seismic) that are consistent with the current code design requirements. These current code-required loads can exceed the original design loads that were used in the existing structure or foundation. But for a replacement in kind no change occurs in the existing risk, therefore an evaluation of the existing structure or foundation is generally not needed.

For replacement in kind, grandfathering the existing foundation or structure to the original design basis is common practice. If the site is under the jurisdiction of a building official, then this approach would be subject to their review and acceptance.

The condition of the existing structural members directly supporting the new replacement-in-kind equipment should always be assessed to confirm that they have not deteriorated.

2.8 CONSIDERATIONS FOR REGULATORY REQUIREMENTS

Prior to construction of a project, whether a new facility or modification to an existing one, submittal to a local jurisdiction for the purpose of obtaining a construction building permit may be required. The current national model building code has many specific requirements for inclusion in the construction documents. In addition, local jurisdictions are likely to have amendments to the adopted national code. A complete permit application package, prepared by a registered design professional, will likely contain design drawings; calculations; and written inspection, observation, and quality assurance plans. The engineer will be required to identify on the construction documents the risk category, occupant load, construction type, fire resistance, egress, and other items specific to the structure. The engineer should identify how those sections of the code apply to the structure in question. The types, concentrations, and amounts of materials in the facility's process determine whether the structure's risk category falls into a hazardous category or a less restrictive one. Construction type and fire resistance will affect the height of the structure, and occupant load affects the egress requirements. Seismic design category is a function of the local site design response spectra and risk category and ranges from "A," the least restrictive, to "F," the most restrictive. The seismic design category will determine whether a given structural system can be used and the resulting height limitations. It also affects whether certain horizontal or vertical irregularities are permitted and the selection of how a structure may be analyzed.

The engineer should carefully consider risk category when classifying a structure in a petrochemical or other industrial facility. Risk category directly influences seismic importance factor, I_e, and seismic design category. Understanding that different structures within a facility can have different importance factors assigned to them is important. Most process facilities likely fall into Risk Category III, with a seismic $I_e = 1.25$ and maximum Seismic Design Category "E" designation. Certain facilities, such as process control rooms that are required to function to properly shut down one or more process units after a major seismic event, may need to be classified as Risk Category IV with a seismic $I_e = 1.5$ and up to a maximum Seismic Design Category "F" designation. The local Authority Having Jurisdiction may require classification as Risk Category IV based on the type and quantities of toxic substances present. Section 1.5.3 of ASCE 7-16 provides, for a building or structure containing toxic, highly toxic, or explosive substances, a possible reduction from a Risk Category IV or III to Risk Category II with $I_e = 1.0$. An owner must demonstrate to the satisfaction of the Authority Having Jurisdiction that a release of any of these substances does not pose a threat to the public. The owner must have performed a hazard assessment as part of an overall risk mitigation plan. The risk mitigation plan must incorporate the hazard assessment, a prevention program, and an emergency response plan. The reader is referred to Section 1.5.3 of ASCE 7-16 for detailed requirements of the three elements of the risk mitigation plan. Section 1.8 of ASCE 7-16 adopts 29 CFR 1910.1200 Appendix A, Health Hazard Definitions (OSHA 2000), as a consensus standard. This OSHA standard's definitions of "highly toxic" and "toxic" substances are identical to the definitions of those terms in IBC Section 307, High-Hazard Group H.

For facilities governed by the IBC, structures and buildings must be classified for use. Quantities of hazardous materials present in a process building or structure in excess of the tabulated limits will require classification in the hazardous group. The specific subgroup classification will depend on whether the process materials present are combustible, flammable, or explosive liquids, gases, and dusts and the range of values for their associated physical properties: flash point, vapor pressures, ignition points, etc. The assigned risk category will govern the construction type and level of fire protection required. In moderate to severe seismic design categories, larger areas and structure heights necessary for the successful function of a given petrochemical or industrial process may be realized by maximizing the fire ratings of structural members. This may govern the selection of fireproofing materials, particularly where weight is a concern, that is, use of proprietary intumescent coating or lightweight proprietary cementitious fireproofing profiled over structural steel as opposed to normal weight box-type concrete fireproofing.

The IBC requires design load conditions to be shown on the construction documents. In addition to dead, live, snow, flood, and wind data, all of the pertinent seismic data are required: importance factor, spectral response factors, site class, seismic design category, base shear, response coefficient, response factor, and identification of the seismic force–resisting system and the type of analysis procedure used. Inclusion of seismic data is required even if seismic loads do not control design.

Importance factors may be different for a structure and for the components of a structure. Note that importance factors for snow loading will be different from seismic loading for the same risk category. Critical piping systems and equipment may require use of component importance factor $I_p = 1.5$, where the supporting structure in Category III would use $I_e = 1.25$. The piping system or equipment would be designed for the higher value and the structure for the lower value. An example where $I_p = I_e = 1.5$ might occur is when a single vertically cantilevered column supports a pipe or piping that requires $I_p = 1.5$. Using the more stringent value for both the supporting structure and the component may be prudent in this instance.

The model building code also has detailed requirements for structural observation and for written seismic force–resisting system special inspection and testing plans to be included in the construction documents for piling, foundations, concrete, steel, and masonry construction. If the structure is in Seismic Design Category "C" or greater, a written seismic quality assurance plan for the seismic force–resisting systems is also required. The plan identifies the components of the seismic force–resisting systems and the types and frequencies of testing, inspection, and observation. As the severity of the seismic design category increases, additional systems must be included in the plan.

2.9 CONSIDERATIONS FOR TEMPORARY FACILITIES

Temporary facilities or structures are considered to be those in service no longer than about 6 months. Some examples of temporary facilities or structures are

- Structures or equipment needed during a construction phase or a plant scheduled shutdown or maintenance period,
- Structures erected to support heavy equipment temporarily prior to locating them on their permanent supporting structures, and
- Structures intended only to support an existing structure while being modified.

Temporary facilities or structures should be completely removed upon the expiration of the stated time limit.

Seismic Design Requirements:
Temporary structures may be designed in accordance with ASCE 37-14, *Design Loads on Structures during Construction,* and ASCE 7 with the following considerations.

- The mapped values for S_S and S_1 may be multiplied by a factor less than 1 to represent the reduced exposure period, but the factor shall not be less than 0.20. The mapped values S_S and S_1 are defined in Chapter 4.

- The response modification coefficient, R, should not exceed 2.5 for bracing systems unless they are detailed in accordance with the provisions of ASCE 7.

2.10 STRUCTURAL OBSERVATION AND INSPECTION

Construction of petrochemical and other industrial facilities in accordance with the explicit or implicit engineering intent is essential for achieving a level of quality that will ensure acceptable behavior of structures and components during ground shaking. Therefore, specific items for which structural observation, inspection, and/or testing is deemed necessary are listed in appropriate sections of this document. Structural observation (site visits) by the engineer of record is recommended to assess general compliance with the design. These site visits are not intended to constitute detailed inspections or to supplant any aspect of required testing and inspection programs.

2.11 QUALITY ASSURANCE

It is recommended that petrochemical and other industrial facilities be designed using an engineering quality assurance plan. Such a plan should include the definition of the earthquake loading to be used in the design and a description of the lateral force–resisting system. Any materials testing program, seismic review procedures, structural observation by the engineer of record, and other inspection programs should be clearly defined.

In evaluation and retrofit of existing facilities, the engineer who identified the need for modification may not necessarily be the engineer who provides the modification design. In such cases, the owner should include review provisions to ensure that the intent of the evaluating engineer is properly implemented. These observations are also intended to ensure that the subsequent construction is carried out properly. Furthermore, the engineer of record should review deviations made in the field to avoid violating the intent of the retrofit.

2.12 PEER REVIEW

New design for hazardous and essential facilities should include an independent peer review. Such independent reviews may be performed by qualified engineers from the design organization who are unrelated to the project. These reviews should examine design philosophy, design criteria, structural systems, construction materials, and other factors pertinent to the seismic capacity of the facility. The review need not provide a detailed check, as normally performed in the design process, but as a minimum should provide an overview to help identify potential oversights, errors, conceptual deficiencies, and other potential problems that might affect a facility's seismic performance.

For new design and for the evaluation of existing facilities, the guidance provided in this document may differ from the current codes. In certain instances, less stringent criteria are recommended in this document. In such cases, an independent peer review is also appropriate.

References

ASCE. 2014. *Design loads on structures during construction.* ASCE 37-14. Reston, VA: ASCE.

ASCE. 2016. *Minimum design loads and associated criteria for buildings and other structures.* ASCE/SEI 7-16. Reston, VA: ASCE.

FEMA. 2015. *National earthquake hazard reduction program recommended provisions for seismic regulations for new buildings and other structures—Part 1. Provisions and Part 2 Commentary.* FEMA 1050. Washington, DC: Building Seismic Safety Council.

IBC (International Building Code). 2018. *International building code.* Country Club Hills, IL: International Code Council.

OSHA. 2000. *Occupational safety and health administration standards.* 29 CFR 1910. Washington, DC: US Dept. of Labor.

CHAPTER 3

Seismic Hazards

3.1 INTRODUCTION

Defining seismic hazards is one of the most important steps in the seismic design and evaluation of petrochemical and other industrial facilities. Engineers who perform seismic design are usually not responsible for the assessment and quantification of seismic hazards; yet, owners and regulators often call on them to interpret the results of hazard analysis and to justify their appropriateness. Therefore, design engineers must clearly understand how specialists in geology, seismology and geotechnical engineering develop this information.

Chapters 11, 16, 20, 21, and 22 of ASCE 7-16 cover assessment of seismic hazards. These chapters provide specific requirements on issues ranging from computation of ground motions for design to requirements of geologic and geotechnical investigations and can sometimes be confusing for people not regularly involved in assessing seismic hazards. This committee believes that a proper understanding of the basics of seismic hazard assessment, especially computation of ground motions, is necessary to develop an appropriate interface among seismologists, geologists, engineers, owners, and regulators involved in the design and evaluation process.

This chapter does not intend to instruct the engineers on how to perform seismic hazard assessment; rather, it provides the necessary background information on techniques used to define and quantify seismic hazards. Readers interested in learning about the mechanics and detailed procedures of seismic hazard analysis for developing design ground motions are referred to *Seismic Hazard and Risk Analysis,* a monograph published by the Earthquake Engineering Research Institute (McGuire 2004) and "Documentation for the 2008 Update of the United States National Seismic Hazard Maps" by the US Geological Survey (Petersen et al. 2008).

Hazards covered in this chapter include ground shaking; excessive ground deformations or ground failure caused by fault rupture, liquefaction, and landslides; and inundation of coastal areas and resulting damage caused by large waves produced in nearby bodies of water (tsunamis and seiches). Appendixes to this chapter provide more detailed discussions of several specific topics.

3.2 EARTHQUAKE BASICS

3.2.1 Earthquake Mechanism

Earthquakes generally result from a sudden release of built-up strain in the earth's crust as slip along geologic faults. Faults are classified on the basis of rupture mechanism and the direction of movement on one side of the fault relative to the other. The three basic types of fault rupture are strike-slip, normal, and reverse/ thrust. Figure 3-1 shows examples of fault slip and its manifestation on the ground surface.

3.2.2 Earthquake Magnitude and Intensity

The size of an earthquake is often described in terms of its magnitude and resulting intensity of ground shaking. The former is a measure of the overall energy released, while the latter is a description of the effects experienced at a particular location during the earthquake.

The magnitude is the single most commonly used descriptor of the size of an earthquake and is typically reported as the "Richter magnitude" in the news media. The Richter magnitude, also known as the local magnitude (M_L), is one of several magnitude scales that use the amplitude of seismic waves to compute the earthquake magnitude. Other such magnitude scales include the surface wave magnitude (M_S) and the body wave magnitude (m_b).

A limitation of such magnitude scales is that the wave amplitude does not continue to increase proportionately with the size of the earthquake. As a result, most magnitude scales reach an upper limit, beyond which the magnitude does not increase commensurate with the size of the earthquake. To avoid this problem, Hanks and Kanamori (1979) introduced the moment magnitude (M_W) scale, which is a function of the total rupture area of the fault and the rigidity of rock rather than the amplitude of seismic waves. The moment magnitude is now the

Strike-Slip Faulting	Reverse Faulting	Normal Faulting
(1906 San Francisco	(1980El Esnam, Algeria Earthquake)	(1983 Borah Peak, Idaho
Earthquake)	Source: NOAA	Earthquake)
Source: USGS		Source: NOAA
Photographer: G. K. Gilbert		

Figure 3-1. Examples of surface fault rupture.
Source: (a) USGS, photographer G. K. Gilbert, (b) NOAA, (c) NOAA.

most commonly used magnitude scale by seismologists. Appendix 3.A presents more detailed descriptions of each of the common magnitude scales and a comparison with the moment magnitude.

Earthquake intensity is a measure of the effects of an earthquake at a particular location. Intensity is a function of the earthquake magnitude, distance from the earthquake source, local soil characteristics, and other parameters (such as the type of damage and human response). The most common measure of intensity in the United States is the Modified Mercalli Intensity (MMI) scale, which ranges from I (not felt) to XII (complete destruction). Appendix 3.A provides a detailed description of the MMI scale. Although common in the United States and many other countries, the MMI scale is not universally used; the Japanese use the Japan Meteorological Agency (JMA) scale, which ranges from 0 to 7, and Eastern European countries use the Medvedev-Sponheuer-Karnik (MSK) scale, with intensities that are similar to the MMI scale.

3.3 GROUND SHAKING

3.3.1 Measures of Ground Shaking for Use in Design and Evaluation

Although the earthquake magnitude and intensity provide a fairly good estimate of earthquake size and its general effects, they are of limited significance in engineering design. As practiced in the industry, engineering design requires earthquake input to be the force or displacement demands on the structure. These demands cannot be estimated directly from either earthquake magnitude or intensity, but can be correlated to the acceleration, velocity, or displacement of the ground at a particular location. Of these, the ground acceleration is directly proportional to the inertial force in the structure and is the most commonly used parameter for design.

During an earthquake, the ground acceleration varies from one location to another. The variation is a function of the nature and size of the earthquake, the proximity of the site to the seismogenic fault, the wave propagation characteristics, and the local subsurface soil conditions. Ground acceleration is typically measured by an accelerometer, which records the amplitude of ground acceleration as a function of time. Typically, two orthogonal horizontal and one vertical components of ground acceleration are recorded. The acceleration time history captures the duration, frequency content, and amplitude of earthquake motions, and it can be integrated with respect to time to compute ground velocity and ground displacement. The recorded acceleration time history is typically digitized over small time increments such as 1/50 s.

Another useful representation of ground shaking is the response spectrum. A response spectrum is the maximum response of a series of single degree of freedom oscillators of mass (m), known period (T), stiffness (K), and constant

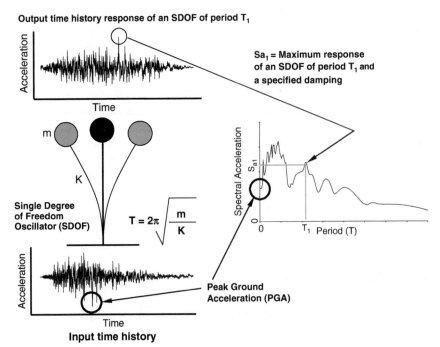

Figure 3-2. Construction of response spectrum.

damping (typically 5%) plotted as a function of their period of vibration (Figure 3-2). The spectral acceleration for a period of zero seconds represents the response of a rigid structure that moves in phase with the ground. This is known as the peak ground acceleration (PGA). The PGA is also the maximum acceleration of the acceleration time history used to compute the response spectrum.

An important distinction should be made between the response spectrum of an acceleration time history and a design response spectrum. The former describes the response of structures of various frequencies to a specific acceleration time history recorded at a site from a unique earthquake, while the latter represents an accumulation of several possible time histories that is usually statistically averaged and smoothed in some manner to define the design demand that the structure is expected to meet in a future earthquake.

3.3.2 Factors Affecting Ground Shaking

The amplitude and duration of strong ground shaking are influenced by the location of the site relative to the fault rupture zone, the size and mechanism of the earthquake, and the subsurface conditions at the site. Unless modified by local subsurface conditions, the amplitude of seismic waves typically decreases with increasing distance from the earthquake source as they propagate through the earth's crust. The decay of ground motion amplitude with distance from the earthquake source is referred to as attenuation.

The common approach for predicting ground motion from an earthquake at a particular site includes utilizing empirically derived ground motion prediction equations (Douglas 2018; EERI 2014), which are also referred to as attenuation relationships. A ground motion prediction equation (GMPE) is a mathematical expression that computes the ordinates of a response spectrum as a function of the earthquake magnitude and its distance from the site. Historically, GMPEs were defined for generic subsurface conditions such as rock or soil; more recently, they are defined for the average shear wave velocity for the top 30 m (approximately 100 ft) of the subsurface (referred to as V_{s30}). Resulting generic ground motions for rock or soil should be modified to account for the local subsurface conditions based on measured or estimated V_{s30} for the site. Other modifications such as basin effects, rupture directivity effects, and source mechanics are also sometime applied, as needed. Because of the uncertainty associated with predicting ground motions and large amount of scatter in the recorded data, a common practice is to use more than one relationship to account for epistemic uncertainty and use a weighted average of the results. Different relationships are available for different tectonic regimes, and the relationship for one regime such as the Western United States should not be applied to the Eastern United States or other parts of the world unless the areas are believed to have similar geologic and tectonic settings. The basic assumptions behind each GMPE must be clearly understood before their application; for example, different relationships use different measures of distance from the earthquake, such as closest distance to rupture or distance to the epicenter, or they have a maximum limit on magnitude and distance for their application.

The following are several general observations regarding ground shaking characteristics:

- The intensity of shaking is proportional to the magnitude of the earthquake. It is also affected by the type of faulting, with thrust-type earthquakes generally producing stronger shaking than strike-slip-type earthquakes.

- For thrust-type earthquakes, the levels of ground shaking could be as much as 20% higher on the up-thrown block or hanging wall than on the down-thrown block or foot wall.

- Deep subduction-zone earthquakes tend to produce lower-frequency and longer-duration ground motions compared with shallow, crustal earthquakes. The lower-frequency-content ground motions from subduction-zone earthquakes tend to increase the response spectrum at longer periods.

- Subsurface conditions influence the levels of ground shaking. With everything else held constant, ground motion at rock has stronger high-frequency content than ground motion at soil. As a result, a rock response spectrum has higher spectral accelerations at short periods than a soil response spectrum, whereas the soil spectrum is higher than the rock spectrum at longer periods (greater than about 0.5 s).

- Motion at the ground surface is influenced mainly by the dynamic characteristics of the top 100 ft of the subsurface soil/rock. Softer sites with lower

V_{s30} have higher spectral accelerations at longer periods of vibration (greater than about 0.5 s). In ASCE 7, subsurface soils are classified in five different categories from hard rock to soft soils (A through E) on the basis of V_{s30} and other geotechnical parameters. ASCE 7 defines a sixth category (F) for soils vulnerable to potential failure or collapse under seismic loading, such as liquefiable soils, quick and highly sensitive clays, and collapsible weakly cemented soils. Chapter 20 of ASCE 7 describes the procedure for site classification using geotechnical parameters or V_{s30}.

- Ground shaking is usually very intense in close proximity to the fault rupture (approximately less than about 10 km), with typically a strong velocity and displacement pulse. This is termed the near-source effect. Within the near-source region, ground shaking in the fault-normal direction is higher than ground shaking in the fault-parallel direction.

- Ground motion is also influenced by rupture directivity. Fault rupture directivity increases the intensity of long-period motions (periods greater than 0.6 s) when the rupture propagates toward the site (forward directivity) and decreases the intensity of motions when it propagates away from the site (backward directivity). This is particularly important for the design of tall or slender structures with long fundamental periods.

3.4 DESIGN GROUND MOTIONS

Assessment of ground motions for design includes either a computation of a design response spectrum or the development of acceleration time histories. Of the two, the former is most commonly used in design. Acceleration time histories, due to extra computational effort, are only used in special cases when a more detailed assessment of structural response is desired or where nonlinear response must be explicitly evaluated.

ASCE 7 provides detailed guidance on the methodology to be used for developing the design response spectrum or acceleration time histories. Chapter 11 of ASCE 7 includes a standard approach for the computation of a design response spectrum. As a substitution to the standard approach, the code allows the use of site-specific analysis (Chapter 21 of ASCE 7) but restricts the resulting spectra to be no less than 80% of the spectra obtained using the standard code approach.

The basic philosophy behind both the standard code and the site-specific approach is to first compute ground motions representative of an extremely rare event and then reduce them by a factor representing the inherent overstrength in a structure to compute design level forces (FEMA P-750 2009). The code refers to this extremely rare event as the maximum considered earthquake (MCE) and specifies a factor of 1.5 as the inherent overstrength. The MCE, as defined in the code, should not be confused by the older and now abandoned acronym for maximum credible earthquake.

In ASCE 7, MCE is defined as ground motions having a 2% probability of exceedance in 50 years. This is generally true for any site within the United States except for sites within close proximity of very active faults, such as coastal California. In these situations, the code allows capping the probabilistically derived 2% in 50-year ground motions by what is considered to be upper bound motions from a single scenario earthquake. The scenario earthquake mainly represents the maximum earthquake that a particular fault could produce (also known as a characteristic earthquake).

ASCE 7 recognizes that the occurrence of such extreme and rare events varies significantly across regions. Using ground motions that represent a uniform hazard as the design basis results in very conservative design in some regions. To provide a consistent level of design conservatism, starting with the 2010 edition, ASCE 7 has introduced risk-targeted MCE (MCE_R) instead of uniform hazard MCE ground motions that were used in previous editions as the design basis. The MCE_R is defined as ground motion that results in 1% probability of collapse within a 50-year period and is obtained by iterative integration of a site-specific ground motion hazard curve (ground motion values versus their annual probability of exceedance) with a fragility curve for collapse (probability of collapse as a function of ground motion).

3.4.1 Design Response Spectrum—Standard Code Approach

The MCE_R ground motions for the standard code approach are based on ground motion maps for the United States developed by the US Geological Survey (USGS). These maps provide contours of spectral acceleration at 0.2 s (also known as short-period spectral acceleration denoted by S_s) and 1.0 s spectral period (denoted by S_1). Spectral accelerations in the code represent maxima of the two orthogonal horizontal directions. Prior to ASCE 7-10, MCE ground motions were based on the geometric mean of the two orthogonal horizontal directions. The MCE_R design parameters (S_s and S_1) for a particular site can be obtained by entering the address or site latitude and longitude coordinates and subsurface soil conditions (in terms of ASCE 7 site class) at the online ASCE 7 Hazard Tool (https://asce7hazardtool.online/).

The MCE_R maps included in ASCE 7 are representative of a generic soft rock site with V_{s30} of 2,500 ft/s (boundary between site classes B and C) and must be adjusted for the actual site conditions if different from the generic condition. The first step in this process is to classify the subsurface conditions in one of the five generic soil classes (A through E) or the sixth soil class F representing liquefiable soils and soils that become unstable in an earthquake. The five generic site classes range from very hard rock "site class A" to very soft soils "site class E" with the intermediate classes B through D representing site conditions between these two extremes.

ASCE 7 site classification is based on specific geotechnical properties of the top 100 ft of the subsurface soil. These properties include average shear wave velocity for the top 100 ft (V_{s30}), average standard penetration test (SPT) blow counts, or average shear strength. Of these, V_{s30} is considered the most reliable and is the preferred approach for site classification. Note that in case of multiple layers of subsurface soils, ASCE 7 includes specific formulas for computing V_{s30}. The formula is based on the assumption of total travel time rather than a direct average of velocity values for each layer. Based on the site class representative of the site conditions, the mapped MCE_R ground motions for V_{s30} of 2,500 ft/s are modified by using site amplification factors F_a and F_v provided in Tables 11.4-1 and 11.4-2 of ASCE 7-16. Factors for site class B through D are based on an estimate of average amplification of the bedrock motions for a range of generic site conditions. Factors for site class E are about 1.3 to 1.4 times the average amplification (refer to Chapter 11 of ASCE 7). If the site conditions are classified as site class F, ASCE 7 (Chapter 21) requires that a detailed site response analysis be performed, which includes analytically modeling the dynamic response of the soil column above bedrock when subjected to seismic shaking at the bedrock level. ASCE 7 also requires the site-specific design response spectrum for site class F to not be less than 80% of the spectrum for site class E. Detailed site-specific ground motion procedures are also required for seismically isolated structures with $S_1 > 0.6g$, structures on site class E sites with $S_s > 1.0g$, and structures on site class D or E sites with $S_1 > 0.2g$. ASCE 7 lists certain conditions for exceptions to the site-specific ground motion requirement.

MCE_R design parameters, after adjusting for local site response, are reduced by a factor of 2/3 (or 1/1.5, the assumed inherent collapse margin for code-designed structures) to obtain the short-period and 1.0 s period spectral ordinates, S_{DS} and S_{D1}. These two design parameters are considered sufficient to construct the entire design response spectrum. This is possible for two key reasons. First, the spectral acceleration (S_a), spectral velocity (S_v), and spectral displacement (S_d) are related to each other as a function of vibration period as shown in Equation (3-1). Second, the general shape of a response spectrum can be idealized into three distinct regions: the acceleration-controlled, the velocity-controlled, and the displacement-controlled (Figure 3-3a). In each of the three regions the respective spectral acceleration, spectral velocity, and spectral displacement are assumed to be constant (Newmark and Hall 1982). On a log-log scale, these regions appear as three straight lines. Response spectrum plotted in this manner is referred to as the tripartite response spectrum (Figure 3-3b).

$$S_a = \frac{2\pi}{T} S_v = \left(\frac{2\pi}{T}\right)^2 S_d \qquad (3\text{-}1)$$

The transition from constant velocity to constant displacement portion of the response spectrum is defined by a transition period (T_L). ASCE 7 provides T_L contour maps for the United States.

3.4.2 Site-Specific Design Response Spectrum

Development of a site-specific design response spectrum includes consideration of regional tectonics (faults and seismic sources, historic seismicity, earthquake

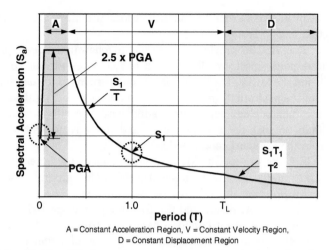

A = Constant Acceleration Region, V = Constant Velocity Region,
D = Constant Displacement Region

Figure 3-3a. Idealized design response spectrum (spectral acceleration versus period—linear scale).

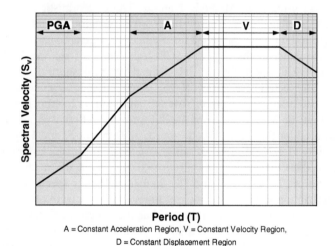

A = Constant Acceleration Region, V = Constant Velocity Region,
D = Constant Displacement Region

Figure 3-3b. Idealized design response spectrum (spectral velocity versus period— log scale).

recurrence rates, etc.), attenuation of ground motion with distance, and the influence of subsurface conditions at the site for which the response spectrum is being computed. Generally speaking, site-specific design response spectra can be computed through either a probabilistic seismic hazard analysis (PSHA) or a deterministic seismic hazard analysis (DSHA). Local site effects in PSHA or DSHA are either considered explicitly through a site response analysis or implicitly by using the appropriate GMPEs. Sections 3.4.2.1 and 3.4.2.2 of this chapter briefly describe PSHA and DSHA procedures, which are discussed in detail in McGuire (2004). Procedures to include local site effects are discussed later in this section and in Section 3.4.2.3.

Chapter 21 of ASCE 7 provides several rules to follow in the development of a site-specific response spectrum. The first set of rules relates to the computation of the MCE and the MCE_R. The second set of rules pertains to the development of design response spectrum from the MCE, and finally, the third set of rules describes the methodology for incorporating local site effects. ASCE 7 also provides limits so that the site-specific ground motions are not significantly different from the standard code-based values.

Per ASCE 7, the site-specific MCE_R can be taken as the lesser of either the probabilistically or the deterministically derived MCE_R ground motions. The probabilistic MCE_R spectral accelerations are computed by either applying risk coefficients (C_{RS} for short period and C_{R1} for 1.0 s period spectral ordinates) to spectral ordinates of 5% damped uniform hazard spectrum with 2% probability of exceedance in 50 years or through iterative integration of a site-specific hazard curve with collapse fragility to obtain motions that result in 1% probability of collapse. Chapter 21 of ASCE 7 refers to the two approaches as Method 1 and Method 2, respectively. The USGS provides web-based calculators for computing MCE_R spectra using both Methods 1 and 2. The deterministic MCE_R ground motions are calculated as 84th percentile 5% damped ground motions from the largest of ground motions computed for characteristic scenario earthquakes on all known active faults in the region. ASCE 7 also imposes a lower limit on the deterministic MCE_R spectrum, which is defined by a peak spectral acceleration of 1.5 g and 1.0 s spectral acceleration of 0.6g. These lower-limit values are for rock and must be adjusted for the appropriate soil conditions at the site. For the lower-limit deterministic MCE_R spectrum ASCE 7 provides different F_a and F_v factors than those used for the standard code approach. Different F_a and F_v factors from those in Tables 11.4-1 and 11.4-2 of ASCE 7-16 for deterministic lower bounds are provided to adjust the shape of the spectrum, which is not accurately represented if the factors from Tables 11.4-1 and 11.4-2 are used. If a site-specific deterministic MCE_R spectrum point falls below the lower-limit spectrum, the lower-limit spectrum point for the same period is used as the design spectrum point, unless the probabilistic spectrum point is lower than the deterministic lower limit spectrum point. If the probabilistic spectrum point is lower than the deterministic spectrum point, then the probabilistic spectrum point is the design spectrum point. This process is repeated for spectrum period points to obtain the full design spectrum.

Once the MCE_R is computed, the design response spectrum is taken as 2/3 of the MCE_R spectrum. The design response spectrum cannot be less than 80% of the spectrum obtained using the regional S_s and S_1 maps included in ASCE 7. However, this restriction may not be applicable for regions that do not have ASCE 7 S_s and S_1 maps, such as locations outside the United States. In addition, the short-period design spectral acceleration (S_{DS}) cannot be less than 90% of the peak spectral acceleration at any period greater than 0.2 s, and the value at 2.0 s cannot be less than half of the 1.0 s parameter (S_{D1}).

3.4.2.1 Probabilistic Seismic Hazard Analysis

In a probabilistic seismic hazard analysis (PSHA), the probability of exceeding a specified level of a ground motion parameter (PGA or spectral acceleration) at a site is computed. The approach accounts for all possible earthquakes with their specified probabilities of occurrence and typically includes simulation of hundreds or thousands of likely earthquake scenarios that could occur anywhere within the site region. Figure 3-4 shows the key elements of PSHA.

Computing a PSHA requires developing a mathematical model of all active and potentially active seismic sources within the site region. This model is known as the seismic source model. Information required for developing a source model includes the location and the geometry of the seismic sources such as specific faults or more generic seismotectonic provinces. Other input parameters include the maximum magnitude potential of these sources and their seismic activity, which is derived either from slip rate or seismicity rate. This information is used with an appropriate GMPE to compute ground motion exceedance probabilities. Appendix 3.A describes various elements of a PSHA.

The results from a PSHA are presented either in terms of uniform hazard spectra for different probabilities of exceedance or a family of curves, known as hazard curves, that show annual exceedance probabilities for each measure of ground motion (such as PGA or spectral acceleration at specified periods of vibration).

Most codes and criteria documents, including ASCE 7, describe the uniform hazard spectra in terms of either return period or ground motion exceedance probability for the design life of a structure rather than the annual probability of exceedance. Assuming a Poisson distribution of hazard, the probability of exceedance and return period are related as given in Equations (3-2) and (3-3).

$$p = 1 - e^{-\lambda t} \tag{3-2}$$

$$\lambda = \frac{1}{T} \tag{3-3}$$

where
   λ = Annual probability of exceedance,
   p = Probability of exceedance in t years, and
   T = Return period.

Commonly, building codes use 2% or 10% probability of exceedance in 50 years. Using Equation (3-2), this translates to a 2,475- or 475-year return period, respectively. The return period as specified here is the return period associated with a level of ground shaking and not the return period (or recurrence interval) of an earthquake. The return period for the maximum event on a given fault may vary from several hundred to several thousand years, depending on the fault's activity rate. Earthquakes can happen any time, or even more than once during this time period.

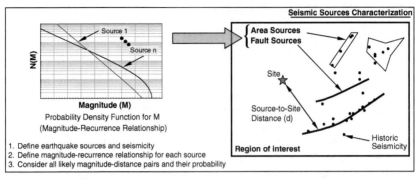

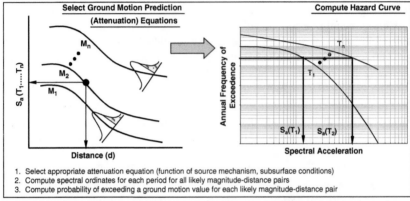

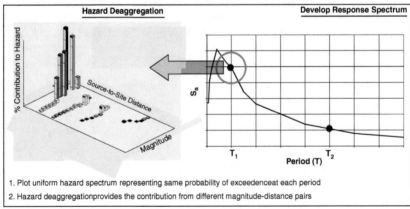

Figure 3-4. Probabilistic seismic hazard analysis.

The hazard curve is typically plotted on a log-log scale and its shape, especially its slope, is strongly influenced by the seismicity of the region. Typically, a seismically active region with frequent earthquakes, such as California, has a relatively flat slope compared with a region with infrequent large earthquakes, such as the New Madrid region, which has a fairly steep slope. The slope of the hazard can significantly influence design ground motions computed using ASCE 7.

In regions with a relatively flat hazard curve, such as California, the two-thirds reduction of the probabilistically derived MCE (2,475-year return period) yields design ground motions with return periods on the order of 475 years, while for regions such as the Central and Eastern United States, with a steep hazard curve, this reduction leads to design ground motions with much longer return periods. Prior to ASCE 7, design ground motions in most building codes were defined as ground motions with a 475-year return period in all regions regardless of the level of seismicity. With the change in the definition of design ground motions (two-thirds MCE), there was a significant increase in these motions in low seismicity regions. Recognizing the potential for overdesign in these regions, starting with the 2010 edition, ASCE 7 uses MCE_R, which generally represents a uniform risk of collapse. The basis for changing the design approach from uniform hazard ground motions to risk-targeted ground motions was developed as part of Project 07, a joint effort of the Building Seismic Safety Council, FEMA, and USGS (Kircher et al. 2010). MCE_R motions are obtained by integrating a standard structural collapse fragility curve as described in FEMA P-695 (FEMA 2009).

3.4.2.2 Deterministic Seismic Hazard Analysis

In a deterministic seismic hazard analysis (DSHA), earthquake ground motions at a site are estimated for specific earthquake scenarios (magnitude and location relative to the site) using appropriate attenuation relationships. Figure 3-5 shows the key elements of a DSHA.

Deterministic analysis does not consider the likelihood of the event or the uncertainty in its magnitude or location but does include the attenuation uncertainty in terms of standard deviation. The uncertainty is typically accounted for by specifying median or median plus one standard deviation ground motions for a specific earthquake scenario. Median ground motions are typically used for standard structures and median plus one standard deviation (84th percentile) for essential facilities. The 84th percentile ground motions are used to define MCE_R ground motions where DSHA procedures are used.

3.4.2.3 Site Response Analysis

In general, three approaches for considering local site effects are available. The first is the direct use of a GMPE that is representative of the subsurface conditions. Some attenuation equations distinguish only between soil and rock, while others provide ground motion as a function of V_{s30} or ASCE 7-type site classes.

The second approach for modeling local site effects is first computing rock outcrop (surface rock) response spectra using a rock GMPE and then modifying the rock spectrum by generic soil amplification factors such as the F_a and F_v factors in Tables 11.4-1 and 11.4-2 of ASCE 7-16 or other published sources such as EPRI (1993) or Stewart and Seyhan (2013).

The third approach for modeling local site effects is through a detailed dynamic site response analysis. Such an analysis is performed when soil condition cannot be reasonably categorized into one of the standard site conditions, or when

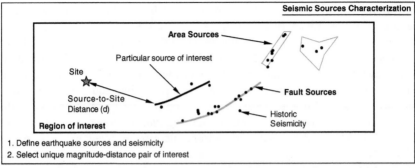

Seismic Sources Characterization

1. Define earthquake sources and seismicity
2. Select unique magnitude-distance pair of interest

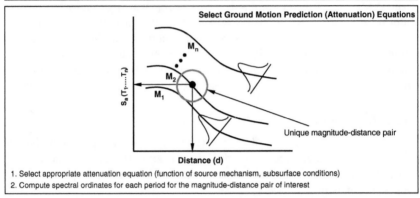

Select Ground Motion Prediction (Attenuation) Equations

1. Select appropriate attenuation equation (function of source mechanism, subsurface conditions)
2. Compute spectral ordinates for each period for the magnitude-distance pair of interest

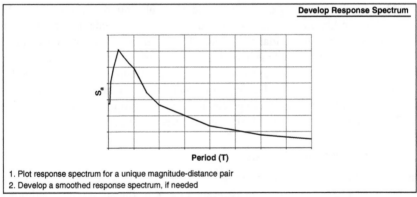

Develop Response Spectrum

1. Plot response spectrum for a unique magnitude-distance pair
2. Develop a smoothed response spectrum, if needed

Figure 3-5. Deterministic seismic hazard analysis.

empirical site factors for the site are not available (such as for site class F). Dynamic site response analysis can either be one-dimensional analysis that assumes vertically propagating shear waves through the various subsurface soil layers. Programs such as SHAKE (Schnabel et al. 1972) or DeepSoil (Hashash et al. 2016) are examples of codes used for one-dimensional equivalent linear analysis, while programs such as FLAC (Itasca Consulting Group, 2008) and OpenSees

(Regents of the University of California 2000) are examples of two-dimensional nonlinear models.

The input to dynamic site response analysis requires identification of subsurface soil strata; specifications of basic and nonlinear soil properties for each layer of subsurface soil, such as unit weight and shear modulus and damping as a function of shear strain (more advanced analysis may require additional soil properties); and specification of input ground motion time histories. Typically, the input time histories for dynamic site response analysis are specified as rock outcrop (rock at ground surface) acceleration time histories that are then modified within the program to represent bedrock (rock at depth) time histories. In cases where the depth to bedrock is very deep and computing the bedrock time histories is impractical, ASCE 7 allows terminating the model at depth where the soil stiffness is at least equivalent to site class D. In such a case, site class D outcrop time histories will be needed for site response analysis. ASCE 7 requires five rock outcrop time histories to be either selected or simulated such that they are representative of earthquake events that control the MCE in terms of magnitude and distance from the fault. ASCE 7 also requires the time histories to be scaled such that on average the response spectrum of each time history is approximately at the level of the MCE rock spectrum over the period range of significance to structural response. This requirement is different and less stringent than the requirement for acceleration time histories that are directly used for structural analysis discussed in the following section.

3.4.3 Earthquake Time Histories

In some cases, earthquake time histories are required as inputs to structural analyses. This may be the case for nonlinear structural analysis and/or analysis of structures of special importance and/or for analysis that considers soil-structure interaction. Chapter 16 of ASCE 7 provides specific requirements for seismic response history analysis that should be followed in selecting appropriate time histories, modifying the selected time histories to meet the design requirements, and applying the time histories in the analysis.

The approach for developing site-specific earthquake time histories should consider the following:

- **Initial selection of time histories**: This includes selecting records that closely match the site tectonic environment, controlling earthquake magnitudes and distances, type of faulting, local site condition, response spectral character-istics, and strong shaking duration. Both recorded time histories from past earthquakes and synthetic time histories can be used. Multiple time histories should be considered. Chapter 16 of ASCE 7requires a minimum of 11 time histories for nonlinear analysis. For seismically isolated structures (ASCE 7, Chapter 17) a minimum of seven time histories are required.

- **Modification of time histories**: Because the selected time histories may differ from the design motions in terms of shaking amplitude and response spectral ordinates, they must be modified for use in analysis. Two modification methods can be used: amplitude scaling or spectral matching.

If amplitude scaling is used, both orthogonal components of the selected time history are linearly scaled using the same scale factor up or down such that the average of the maximum-direction spectra from all selected ground motions generally matches or exceeds the target spectrum and that the average spectrum is not less than 90% for any period within the period range of interest of the target response spectrum. If spectral matching is used, the time history is adjusted either in the frequency domain by varying the amplitudes of the Fourier Amplitude Spectrum on the time domain by adding wavelets in iterations until a satisfactory match to the target spectrum is obtained. According to ASCE 7, the maximum-direction spectra of all-time histories considered in spectral matching should be equal to or exceed 110% of target spectra within the period range of interest.

The period range of interest for amplitude scaling or spectral matching should be at least two times the largest first-mode period and not less than the period that includes 90% of mass participation in each principal horizontal direction. The lower-bound period should not be less than 20% of the smallest first-mode period in each principal direction. ASCE 7 allows a reduction from two times the upper-bound period to 1.5 times if justified by dynamic analysis. For vertical response the lower-bound period need not be less than 0.1 s or the lowest period with significant vertical mass participation.

A uniform hazard response spectrum (in which each ordinate of the spectrum has the same probability of exceedance) is inherently a conservative target for scaling ground motion, because it represents ground motions for a given hazard level and does not necessarily represent actual ground motions. The conditional mean spectrum (CMS) provides an alternative response spectrum for selecting ground motions (Baker 2011). A CMS is appropriate for period-specific ground motions, and multiple CMS spectra targeting different periods may be necessary to fully capture the structural response in different earthquake scenarios.

3.5 GROUND FAILURE

In addition to ground shaking, seismically induced ground failures (geologic hazards) must be evaluated. These hazards include surface fault rupture, liquefaction, and landslides. Ground failures are important concerns for the design and evaluation of petrochemical and other industrial facilities. An unexpected ground failure often proves to be catastrophic, in terms of damage not only to the main structures and equipment but also to buried systems, such as pipelines.

3.5.1 Surface Fault Rupture

Surface rupture is a direct shearing at a site during an earthquake event. It is generally due to moderate to large earthquakes. The offsets or displacements typically have both horizontal and vertical components, and they can be as large as several meters. A ground displacement of more than a few inches has been observed to cause major damage to structures located on the fault. In general, without detailed geologic investigations, the precise location of the fault is not known with a high degree of certainty (within a few hundred feet), and because fault displacements produce large forces and movements, the best way to limit damage to structures is to avoid building in areas close to active faults. Empirical relationships, such as that of Wells and Coppersmith (1994), are commonly used to estimate the magnitudes and distribution of fault rupture displacement and provide maximum and average fault displacements as a function of earthquake magnitude and rupture dimensions. In California, Earthquake Fault Zone maps (formerly known as the Alquist–Priolo maps) are available for major faults. These maps are available online as the California Earthquake Hazards Zone Application ("EQ Zapp") that shows the zones of potential surface faulting and other earthquake hazards such as liquefaction and landslide (https://maps.conservation. ca.gov/cgs/EQZApp/app/).

3.5.2 Liquefaction

Intense shaking from an earthquake can increase the pore water pressure within saturated soils. If the pore pressure reaches the overburden stress, the effective stress between soil grains becomes zero and the soil liquefies. Liquefied soil is extremely soft, has very low strength, and is easily deformed. The temporary reduction in soil strength from liquefaction can result in foundation-bearing failure, ground cracking, and large ground deformations. The elevated pore pressure in liquefied soils can cause uplift of buried tanks, vaults, and pipelines.

Not all soils are susceptible to liquefaction. Assessing the susceptibility of soils to liquefaction is the first step in liquefaction evaluation. Soil liquefaction tends to occur primarily in saturated loose sandy or silty soils. EERI (1994) notes that liquefaction often occurs in areas where the groundwater table is within 30 ft (10 m) of the ground surface, and only a few instances of liquefaction have been observed where groundwater is deeper than 60 ft (20 m). Soils must be saturated and have little or no plasticity to be susceptible to liquefaction. For this reason, soils may be screened for liquefaction susceptibility based on their depth relative to the groundwater table and their Plasticity Index (PI). The National Academies of Science, Engineering, and Medicine Report (2016) provides details about screening soils for liquefaction susceptibility.

The second step in liquefaction evaluation is an assessment of liquefaction triggered by the anticipated intensity of ground shaking. Factors affecting liquefaction potential include dynamic shear stresses in the soil; earthquake magnitude or shaking duration; and soil properties including type, density, gradation, geologic age, and depositional environment. Based on observations from the 2010–2011 Canterbury Earthquake Sequence (CES) in New Zealand, areas with potentially liquefiable soils can repeatedly liquefy in subsequent earthquakes that are large enough to exceed the liquefaction threshold.

The most commonly used analytical procedures to assess the potential for liquefaction triggering is known as the "Simplified Procedure" (Seed and Idriss 1982). The procedure uses two variables, the cyclic stress ratio (CSR) and the cyclic resistance ratio (CRR), for this purpose. CSR characterizes the seismic demand on the soil and is a function of depth, soil composition, effective and total overburden stress, earthquake magnitude (M), and PGA. CRR characterizes the soil's capacity to resist liquefaction and is estimated using soil properties such as SPT blow count values or cone penetrometer test (CPT) cone-tip resistance. CRR is corrected for factors such as overburden stress, hammer energy, and fines content. The factor of safety against liquefaction is computed as a ratio of CRR to CSR. A factor of safety less than 1.0 indicates liquefaction is likely.

The procedures to evaluate liquefaction triggering were developed, in part, based on empirical observations of site performances (i.e., liquefied and non-liquefied sites) in past earthquakes. Approaches for liquefaction assessment include Seed and Idriss (1982), Seed et al. (1983, 1985), Robertson and Wride (1997), Youd et al. (2001), Idriss and Boulanger (2012), and Cetin et al. (2004). Boulanger and Idriss (2014) present the latest revisions and refinements to the procedure and include high-quality liquefaction case histories from recent earthquakes in New Zealand (2010–2011 CES) and Japan (2011 M_w 9.0 Tohoku).

ASCE 7 requires the potential for liquefaction to be evaluated based on the geometric mean PGA from the maximum considered earthquake (MCE_G). The PGA from the MCE_G differs from the PGA calculated using the risk-targeted MCE_R. The PGA for liquefaction analysis should also reflect local site conditions, and a site amplification factor, F_{PGA}, which is based on the ASCE 7 site class, should be applied to calculate the PGA. MCE_G design parameters can be obtained by entering the address or latitude and longitude coordinates of the site and subsurface soil conditions (in terms of ASCE 7 site class) at the online ASCE 7 Hazard Tool (https://asce7hazardtool.online/).

The third step in a liquefaction evaluation is an assessment of the consequences of liquefaction. Liquefaction is typically manifested in terms of bearing-capacity failure, soil settlement, and lateral spreading. Other consequences of liquefaction include ground motion modification, uplift of buried structures, increased lateral loads on embedded structures, and down-drag of pile-supported structures. The following sections discuss some of the potential consequences of liquefaction.

3.5.2.1 Bearing Failure

The material behavior of liquefied soil is very complex, as its stiffness can vary by many orders of magnitude throughout each cycle of loading. At very large shear strains, the liquefied soil may approach the soil's residual shear strength. Instabilities, such as bearing-capacity failure, occur when the static shear stress in the soil exceeds its shear strength. Therefore, the residual shear strength of liquefied soil is an important consideration in the assessment of foundation bearing capacity.

Procedures to estimate the residual shear strength of liquefied soil are most often backcalculated from observations of ground failure in past earthquakes.

Olson and Stark (2002), Idriss and Boulanger (2008), and Robertson (2010) develop estimates of soil residual strength based on empirical observations from past earthquakes. Kramer and Wang (2015) present a model that is based both on critical state soil mechanics and empirical observations. As with estimates of ground motion intensity measures, using a weighted average of an ensemble of residual shear strength estimates is common practice.

3.5.2.2 Settlement

Soil settlement can result from soil densification, excess pore water pressure dissipation following liquefaction, lateral squeezing of liquefied soil under foundation loads, and loss of volume from surface ejecta. Shaking from an earthquake can cause unsaturated soils to compact and decrease in volume (volumetric strain). The vertical accumulation of volumetric strain is reflected in post-earthquake settlement of the ground surface. Volumetric strain also occurs as excess pore water pressure drains from liquefied soil. The magnitude of volumetric strain of liquefied soil is typically larger than the volumetric strain caused by compaction of unsaturated soil. Considerations should be given to identifying the lateral and vertical extents of liquefiable soils so that the potential for nonuniform (differential) settlements can be evaluated. The procedures proposed by Tokimatsu and Seed (1987), Ishihara and Yoshimine (1992), or Wu and Seed (2004) can be used to estimate the amount of settlement as a function of earthquake shaking and SPT blow counts.

Settlement estimates based on volumetric strains do not account for material loss from ejecta (such as sand boils). For example, the soil may relieve excess pore water pressure by ejecting groundwater to the surface, carrying with it fine sands and silts, which are deposited as sand boils. Settlement caused by ejecta is difficult to predict and depends on the depth to the water table; the ratio of nonliquefiable crust thickness to the thickness of the underlying liquefiable soil; and existing fractures, root holes, and structural penetrations, all of which are highly variable. Van Ballegooy et al. (2014) discuss recent experience with settlements caused by ejecta during the Canterbury Earthquake Sequence in New Zealand.

3.5.2.3 Lateral Spreading

Seismically induced horizontal movements can also occur on sloping or gently sloping ground when the underlying soils liquefy. The horizontal movement of a slope underlain by liquefied soil is termed lateral spreading. Lateral spreading can also be initiated at free faces where lateral movement can take place. Case histories (Zhang et al. 2004) indicate that the length of lateral spreading can extend as much as 40 times the height of the free face. Typical procedures for lateral spread assessment include empirical methods (Khoshnevisan et al. 2015).

Very often, the complicated behavior of liquefied soil can only be appropriately evaluated using advanced, nonlinear effective stress models in programs such as FLAC or OpenSees. These models can also assess the dynamic kinematic loads on structures caused by laterally deforming soils. However, numerical modeling is a significant effort and requires expertise and experience.

3.5.2.4 Liquefaction Mitigation Measures

Mitigation measures for liquefaction vary from removing liquefiable soils to strengthening foundation systems to resist liquefaction. Table 3-1 (Seed et al. 2003) lists some of the more common mitigation methods. The selection of one or more methods should be evaluated based on constructability, effectiveness, cost, and other project constraints, such as environmental impacts and permitting and project schedule. Consideration should also be given to the ability to verify the selected methods' effectiveness. For example, if stone columns are used to densify liquefiable soils, SPT blow counts can be measured before and after stone column installation so that their effectiveness can be verified. Section 12.13.9 of ASCE 7-16 provides new foundation requirements for building structures located on liquefiable sites.

3.5.3 Landslides

Seismically induced, or coseismic, landslides may encompass mass volume and area that are much larger than those due to other causes. Sometimes, these landslides involve areas of many square miles and may be located more than 100 miles from the epicenter. Most coseismic landslides have been triggered by seismic events greater than magnitude 5.

Landslides can occur in slopes with and without the presence of liquefied soils. Both soil strength loss and inertial loading due to shaking can contribute to coseismic landslide deformation. In conditions where soil strength is not expected to change significantly during shaking, pseudo-static analyses can be used to estimate a yield acceleration, which is the acceleration that would temporarily destabilize the slope. The pseudo-static yield acceleration can then be used in conjunction with sliding block regression analyses, such as Bray and Travasarou (2007) or Kim and Sitar (2004), to estimate coseismic landslide deformation. Advanced numerical finite element or finite difference methods are necessary to evaluate the dynamic slope responses when the soil strength degrades significantly during shaking.

3.6 TSUNAMI AND SEICHE

Tsunamis and seiches occur regularly throughout the world, but are usually only considered in the design and evaluation of facilities when making initial decisions on the facility's location. This section briefly describes these phenomena, assessment methods, and possible mitigation measures. Appendix 3.B provides additional discussion.

Tsunamis are typically generated by large and sometimes distant earthquakes associated with undersea faulting, submarine landsliding, or volcanic eruption. Traveling through the deep ocean, a tsunami is a broad and shallow, but fast-moving, wave that poses little danger to most vessels in deep water. When

Table 3-1. List of Methods for Mitigation of Soil Liquefaction.

General Category	Mitigation Methods	Notes
I. Excavation and/or compaction	Excavation and disposal of liquefiable soils Excavation and recompaction Compaction (for new fill)	
II. In situ ground densification	Compaction with vibratory probes (e.g., Vibroflotation, Terraprobe, etc.)	Can be coupled with the installation of gravel columns
	Dynamic consolidation (Heavy tamping) Compaction piles/stone columns	Can also provide reinforcement
	Deep densification by blasting Compaction grouting	
III. Selected other types of ground treatment	Permeation grouting Jet grouting Deep mixing	
	Drains: gravel drains, sand drains, prefabricated strip drains	Many drain installation processes also provide in situ densification
IV. Berms, dikes, seawalls, and other edge containment structures/systems	Structures and/or earth structures built to provide edge containment and thus to prevent larger lateral spreading	
V. Deep foundations	Piles (installed by driving or vibration) Piers (installed by drilling or excavation)	Can also provide ground densification
VI. Reinforced shallow foundations	Grade beams Reinforced mat Well-reinforced and/or post-tensioned mat "Rigid" raft	

tsunamis reach the coastline, wavelengths become shortened and wave amplitudes increase. Coastal waters may rise above normal sea level and wash inland with great force. The succeeding outflow of water is just as destructive as the tsunami run-up.

A seiche occurs when resonant wave oscillations form in an enclosed or semi-enclosed body of water such as a lake, bay, or fjord. Seiches are typically caused when strong winds and rapid changes in atmospheric pressure push water from one end of a body of water to the other. The water then rebounds to the other side of the enclosed water body and can oscillate back and forth for time periods ranging from a few minutes to several days. Earthquakes, tsunamis, or severe storm fronts may also cause seiches along ocean shelves and ocean harbors.

A tsunami or seiche may result in rapid flooding of low-lying coastal areas. The greatest hazard results from the inflow and outflow of water, where strong currents and forces can erode foundations and sweep away structures, vehicles, vessels, or almost any large body in its path. Petrochemical facilities are especially vulnerable to the rupture or movement of storage tanks from debris impact, foundation erosion, or buoyancy, and can result in massive pollution, fires, or explosions. The tsunami at Seward, Alaska, following the 1964 Alaska Earthquake led to destruction of port facilities. The succession of waves spread the fire from ruptured oil tanks across and throughout the port area. The flooding of low-lying coastal areas from the December 26, 2004, tsunami in Southeast Asia resulted in one of the most devastating natural disasters of the past 100 years, with more than 220,000 deaths and about 450,000 displaced homeless survivors. As a result of the 2004 tsunami, the petroleum storage and distribution facility at the deep-water port at Kreung Raya lost half of its above-ground piping and three of 12 liquid fuel storage tanks. None of the tanks was anchored to its foundations, and the three that were swept away by tsunami waves were only partially full.

Tsunamis can cause more damage than strong ground shaking from an earthquake. The 2011 tsunami in Japan overturned petrochemical storage tanks as a result of wave impact and floating, often accompanied by the release of significant quantities of hydrocarbons. Overall, 1,404 hazardous facilities were damaged by the M9.0 earthquake and 1,807 by the subsequent tsunami. Production facilities at Japan's largest petrochemical complex, in Kashima, about 50 mi northeast of Tokyo, were flooded with up to three ft of seawater by the 2011 tsunami, resulting in the loss of at least 1.7 million t/yr of ethylene capacity and more than 1 million bbl/day of crude processing capacity.

A third potential cause of coastal inundation is coastal subsidence caused by tectonic (faulting) or nontectonic (e.g., submarine landslide) effects. Permanent coastal subsidence and submergence may occur. Predicting when, or if, such tectonic subsidence will occur in a given area is difficult, if not impossible, although historical earthquake data provide evidence of such occurrences. With a nearby submarine slide, there may be no time for warning or escape.

Coastal sites along active tectonic subduction zones where dip-slip faulting is common are most susceptible to earthquake-induced inundation. Examination of the regional fault characteristics and earthquake history, with emphasis on submarine earthquakes and possible historical tsunamis, provides the best measure of site vulnerability. Estimates of maximum tsunami run-up may be based on either historical occurrences or theoretical modeling.

In Japan, where tsunamis are common, several mitigation measures have been implemented, such as sea walls and barriers to resist inundation. These structures are sometimes unsuccessful, as large tsunamis may overtop the barriers and flood the protected area, generating strong currents that erode the barriers and other structures. Many people who were caught in March 11, 2011, Tōhoku Earthquake and tsunami that resulted in 15,895 deaths and 2,539 missing thought they were safe behind tsunami walls that had been constructed to protect against tsunamis of much lower heights. Coastal protective structures must be carefully designed to withstand extreme events. These protective structures can provide a false sense of security or end up acting as powerful battering rams against the very structures they were designed to protect.

Petrochemical facilities located along coastlines may be subject to tsunami inundation and possible damage. Facilities should have a tsunami plan in place, with specific actions to be taken upon notification of a tsunami. If the tsunami is from a distant source, facilities may have hours to prepare and shut down operations. A nearshore tsunami may only give a 5 to 10 min warning and require a different action plan. Chapter 6 of ASCE 7 and its commentary provide detailed guidelines for computing tsunami loads, inundation levels and mitigation measures.

References

ASCE. 2016. *Minimum design loads and associated criteria for buildings and other structures*. ASCE/SEI 7-16. Reston, VA: ASCE.

Baker, J. W. 2011. "Conditional mean spectrum: Tool for ground motion selection." *J. Struct. Eng.* **137** (3): 322–331.

Bolt, B. A. 1988. *Earthquakes*. New York: W. H. Freeman.

Boulanger, R. W., and I. M. Idriss. 2014. *CPT and SPT based liquefaction triggering procedures*. Rep. No. UCD/CGM-14/01. Davis, CA: Univ. of California Center for Geotechnical Modeling, Dept. of Civil and Environmental Engineering, College of Engineering.

Bray, J. D., and T. Travasarou. 2007. "Simplified procedure for estimating earthquake-induced deviatoric slope displacements." *J. Geotech. Geoenviron. Eng.* **133** (4): 381–392.

Cetin, K. O., R. B. Seed, A. Der Kiureghian, K. Tokimatsu, L. F. Harder, R. E. Kayen, et al. 2004. "Standard penetration test-based probabilistic and deterministic assessment of seismic soil liquefaction potential." *J. Geotech. Geoenviron. Eng.* **130** (12): 1314–1340.

Cornell, C. A. 1968. "Engineering seismic risk analysis." *Bull. Seismol. Soc. Am.* **58** (5): 1583–1605.

Cornell, C. A., and E. H. Vanmarcke. 1969. "The major influences of seismic risk." In Vol. **1** of *Proc., 4th World Conf. on Earthquake Engineering*, Santiago, Chile, 69–83.

Der Kiureghian, A., and A. H.-S. Ang. 1977. "A fault-rupture model for seismic risk analysis." *Bull. Seismol. Soc. Am.* **67** (4): 1173–1194.

Douglas, J. 2018. *Ground motion prediction equations 1964–2018*. Glasgow, UK: Univ. of Strathclyde, Dept. of Civil and Environmental Engineering.

EERI (Earthquake Engineering Research Institute). 1994. *Liquefaction: Earthquake basics brief no. 1*. Oakland, CA: EERI.

EERI. 2014. "NGA-West2 Research Project." *Earthquake spectra* **30** (3): 973–987.

EPRI (Electric Power Research Institute). 1993. *Guidelines for determining design basis ground motions*. EPRI TR-102293. Palo Alto, CA: EPRI.

FEMA. 2009a. *NEHRP recommended seismic provisions for new buildings and other structures*. FEMA P-750. Washington, DC: FEMA.

FEMA. 2009b. *Quantification of building seismic performance factors*. FEMA P-695. Washington, DC: FEMA.

Gutenberg, B., and C. F. Richter. 1956. "Earthquake magnitude, intensity, energy, and acceleration." *Bull. Seismol. Soc. Am.* **46** (2): 105–145.

Hanks, T. C., and H. Kanamori. 1979. "A moment magnitude scale." *J. Geophys. Res.* **84** (B5): 2348–2350.

Hashash, Y. M. A., M. I. Musgrove, J. A. Harmon, D. R. Groholski, C. A. Phillips, and D. Park. 2016. *DEEPSOIL 6.1, user manual*. Urbana, IL: Board of Trustees of Univ. of Illinois at Urbana-Champaign.

Houston, J. R., and A. W. Garcia. 1978. *Type 16 flood insurance study: Tsunami predictions for the west coast of the Continental United States*. Technical Rep. No. H-78-26. Alexandria, VA: USACE Waterways Experiment Station.

Idriss, I. M., and R. W. Boulanger. 2008. *Soil liquefaction during earthquakes*. Monograph MNO-12. Oakland, CA: Earthquake Engineering Research Institute.

Idriss, I. M., and R. W. Boulanger. 2012. "Examination of SPT-based liquefaction triggering correlations." *Earthquake Spectra* **28** (3): 989–1018.

Ishihara, K., and M. Yoshimine. 1992. "Evaluation of settlements in sand deposits following liquefaction during earthquakes." *Soils Found.* **32** (1): 173–188.

Itasca Consulting Group. 2008. *Fast Lagrangian analysis of continua, user's manual, FLAC version 6.0*. Minneapolis: Itasca Consulting Group.

Kim, J., and N. Sitar. 2004. "Direct estimation of yield acceleration in slope stability analyses." *J. Geotech. Geoenviron. Eng.* **130** (1): 111–115.

Kircher, C. A., N. Luco, and A. S. Whittaker. 2010. "Project 07—Reassessment of seismic design procedures." In *Proc., 2010 Structures Congress*, Orlando, FL.

Khoshnevisan, S., H. Juang, Y. Zhou, and W. Gong. 2015. "Probabilistic assessment of liquefaction-induced lateral spreads using CPT—Focusing on the 2010–2011 Canterbury earthquake sequence." *Eng. Geol.* **192** (Apr): 113–128.

Kramer, S. L., and C. Wang. 2015. "Empirical model for estimation of the residual strength of liquefied soil." *J. Geotech. Geoenviron. Eng.* **141** (9): 04015038.

Merz, H. A., and C. A. Cornell. 1973. "Seismic risk analysis based on a quadratic frequency model." *Bull. Seismol. Soc. Am.* **63** (6): 1999–2006.

McCulloch, D. S. 1985. "Evaluating tsunami potential." In *Evaluating earthquake hazards in the Los Angeles region: US geological survey professional paper 1360*, J. I. Ziony, ed., 375–413. Reston, VA: USGS.

McGuire, R. K. 2004. *Seismic hazard and risk analysis*. Oakland, CA: Earthquake Engineering Research Institute.

NASEM (National Academies of Science, Engineering, and Medicine). 2016. *State of the art and practice in the assessment of earthquake-induced soil liquefaction and its consequences*. Washington, DC: NASEM.

Newmark, N. W., and W. J. Hall. 1982. *Earthquake spectra and design*. Oakland, CA: Earthquake Engineering Research Institute.

Olson, S. M., and T. D. Stark. 2002. "Liquefied strength ratio from liquefaction flow failure case histories." *Can. Geotech. J.* **39** (3): 629–647.

Petersen, M. D., A. D. Frankel, S. C. Harmsen, C. S. Mueller, K. M. Haller, R. L. Wheeler, et al. 2008. *Documentation for the 2008 update of the United States National Seismic Hazard Maps.* USGS Open-File Rep. No. 2008-1128. Reston, VA: USGS.

Power, M., B. Chiou, N. Abrahamson, Y. Bozorgnia, T. Shantz, and C. Roblee. 2008. "An overview of the NGA project." *Earthquake Spectra* **24** (1): 3–21.

Regents of the University of California. 2000. *Open system for earthquake engineering simulation (OpenSees).* Berkeley, CA: Pacific Earthquake Engineering Research Center.

Reiter, L. 1990. *Earthquake hazard analysis issues and insights.* New York: Columbia University Press.

Richter, C. F. 1958. *Elementary seismology.* San Francisco: W. H. Freeman.

Robertson, P. K. 2010. "Evaluation of flow liquefaction and liquefied strength using the cone penetration test." *J. Geotech. Geoenviron. Eng.* **136** (Jun): 842–853.

Robertson, P. K., and C. E. Wride. 1997. "Cyclic liquefaction and its evaluation based on SPTand CPT." In *Proc., Seismic Short Course on Evaluation and Mitigation of Earthquake Induced Liquefaction Hazards, NCEER Workshop,* San Francisco.

Schnabel, P. N., J. Lysmer, and H. B. Seed. 1972. *SHAKE, a computer program for earthquake response analysis of horizontally layered sites.* EERC Rep. No. 72-12. Berkeley, CA: Univ. of California Earthquake Engineering Research Center.

Seed, H. B., and I. M. Idriss. 1982. Vol. **5** of *Ground motions and soil liquefaction during earthquakes.* Engineering Monograph. Oakland, CA: Earthquake Engineering Research Institute.

Seed, H. B., I. M. Idriss, and I. Arango. 1983. "Evaluation of liquefaction potential using field performance data." *J. Geotech. Eng.* **109** (3): 458–482.

Seed, H. B., K. Tokimatsu, L. F. Harder, and R. M. Chung. 1985. *The influence of SPT procedures in soil liquefaction resistance evaluations.* Rep. No. UCB/EERC-84/15. Berkeley, CA: Univ. of California Earthquake Engineering Research Center.

Seed, R. B., K. O. Cetin, R. E. S. Moss, A. M. Kammerer, J. Wu, J. M. Pestana, et al. 2003. "Recent advances in soil liquefaction engineering: A unified and consistent framework." In *Proc., 26th Annual ASCE Los Angeles Geotechnical Spring Seminar, Keynote Presentation.* Long Beach, CA.

Stewart, J. P., and E. Seyhan. 2013. *Semi-empirical nonlinear site amplification and its application in NEHRP site factors.* PEER Rep. No. 2013. Berkeley, CA: Univ. of California Pacific Earthquake Engineering Research Center.

Tokimatsu, K., and H. B. Seed. 1987. "Evaluation of settlements in sand due to earthquake shaking." *J. Geotech. Eng.* **113** (8): 861–878.

UBC (Uniform Building Code). 1997. *International conference of building officials.* Whittier, CA: UBC.

van Ballegooy, S., S. C. Cox, R. Agnihotri, T. Reynolds, C. Thurlow, H. K. Rutter, et al. 2013. *Median water table elevation in Christchurch and surrounding area after the 4 September 2010 Darfield Earthquake Version 1.* GNS Science Rep. No. 2013/01. Lower Hutt, New Zealand: GNS Science.

Wells, D. L., and K. J. Coppersmith. 1994. "New empirical relationships among magnitude, rupture length, rupture width, rupture area, and surface displacement." *Bull. Seismol. Soc. Am.* **84** (4): 974–1002.

Wu, J., and R. B. Seed. 2004. "Estimation of liquefaction-induced ground settlement (case studies)." In *Proc., 5th Int. Conf. on Case Histories in Geotechnical Engineering, Paper No. 3.09.* New York.

Youd, T. L., I. M. Idriss, R. D. Andrus, I. Arango, G. Castro, J. T. Christian, et al. 2001. "Liquefaction resistance of soils: Summary report from the 1996 NCEER and 1998 NCEER/NSF workshops on evaluation of liquefaction resistance of soils." *J. Geotech. Geoenviron. Eng.* **127** (10): 817–833.

Youngs, R. R., and K. J. Coppersmith. 1985. "Implications of fault slip rates and earthquake recurrence models to probabilistic seismic hazard estimates." *Bull. Seismol. Soc. Am.* **75** (4): 939–964.

Zhang, G., P. K. Robertson, and R. W. I. Brachman. 2004. "Estimating liquefaction-induced lateral displacements using the standard penetration test or cone penetration test." *J. Geotech. Geoenviron. Eng.* **130** (8): 861–871.

APPENDIX 3.A GROUND SHAKING

3.A.1 INTRODUCTION

The most widely used seismic input parameter in the seismic design of structures is a measure of the ground shaking, usually given in terms of ground acceleration. It is often the only parameter used for specification of seismic design.

Section 3.2 provides background information on common terminology and fundamentals related to the evaluation of ground-shaking hazards. This Appendix provides additional discussion on specific areas that are often misunderstood, including terminology and interpretation and the application of its results. This section aims to explain common ground motion descriptions and their relevance to engineering applications.

3.A.2 EARTHQUAKE BASICS

Earthquakes are generally caused by the sudden release of built-up elastic strain in the earth's crust and originate as slip along geologic faults. The total energy of an earthquake is released along the entire length of the ruptured zone, propagating from the source and traveling through the earth in the form of seismic waves. The point in the earth's crust where the rupture is initiated is known as the focus or the hypocenter of the earthquake. The point on the surface of earth directly above the hypocenter is known as the epicenter of the earthquake.

The theory of plate tectonics can explain the build-up of elastic strain in the earth's crust, positing that the outer 70–150 km of the earth's crust comprises approximately 12 major plates that are slowly moving relative to each other. Most of the seismically active areas are located along the plate boundaries. The relative motion between the plates results in either plates grinding past each other, a plate subducting beneath another, or several plates converging and crushing smaller plates. There are also divergent plate boundaries. Mountains and deep sea trenches are formed where plates converge and subduct beneath one another. Mid-ocean ridges are formed at divergent plate boundaries. Earthquakes usually initiate at shallow depths at plate boundaries where plates slide past each other. Deep focus earthquakes usually occur along subduction plates. Earthquakes within a plate can also result for other reasons such as increased compressional stress. Other causes of earthquakes include nuclear explosions or large artificial reservoirs that change the local state of stress within the earth's crust. Also, volcanic activity can cause earthquakes at both subducting and diverging plate boundaries.

3.A.3 MAGNITUDE AND INTENSITY

The amount of energy released in an earthquake and its size characterize the magnitude of the earthquake. Some of the most commonly used magnitude scales are Richter or local magnitude (M_L), surface wave magnitude (M_s), body wave magnitude (m_b), and moment magnitude (M_w).

Gutenberg and Richter (1956) show that the amount of energy released in an earthquake can be related to the magnitude by the following relationship:

$$\log E = 1.5 M_S + 11.8 \qquad (3.A\text{-}1)$$

where E is the energy in ergs. A magnitude increase by one unit releases 31.6 times more energy.

Richter (1958) defined the magnitude of a local earthquake (M_L) as the "logarithm to base ten of the maximum seismic-wave amplitude (in thousandth of a millimeter) recorded on a standard seismograph at a distance of 100 kilometers from the earthquake epicenter" (Bolt 1988). Seismograms from deep focus earthquakes differ significantly from those for shallow focus earthquakes, even though the total amount of energy released may be the same. This results in different magnitude estimates for earthquakes that release the same amount of energy. This is a limitation of M_L.

The magnitude of an earthquake is also determined by the amplitude of the compressional or P wave. The focal depth of an earthquake does not affect the P wave's amplitude. The magnitude estimated by measuring the P wave amplitude is called the P wave or body wave magnitude (m_b).

Magnitude calculated by measuring the amplitude of long-period surface waves (with periods near 20 s) is called the surface wave magnitude (M_s). Surface wave magnitude is calculated only for shallow focus earthquakes that give rise to surface waves.

For small or moderate size earthquakes, the amplitude of seismic waves measured by typical or standard seismographs increases as the size of the earthquake increases. This trend, however, does not continue for large or very large earthquakes for which the amplitude of seismic waves, whose wave lengths are much smaller than the earthquake source, do not increase proportionally to the size of the earthquake. The magnitude estimated by the amplitude of seismic waves, therefore, does not continue to increase at the same rate as the size of the earthquake. Beyond a certain limit the magnitudes calculated in this way tend to remain constant for different size earthquakes. In other words, the magnitude scale saturates, resulting in no further increase in estimated magnitude with the increasing size of an earthquake.

Because of the limitations in the previously defined magnitude scales, Hanks and Kanamori (1979) introduce a magnitude scale based upon the amount of energy released, which is termed the moment magnitude scale (M_w). Moment magnitude is defined by the following relationship:

$$M_w = \frac{2}{3} \log M_o - 10.7 \qquad (3.A\text{-}2)$$

where

$$M_o = \mu\, AD \qquad (3.A\text{-}3)$$

The seismic moment M_o is defined as a product of the rock's modulus of rigidity (μ), area of rupture (A), and average fault displacement (D) in a seismic event.

M_W, therefore, depends on the size of rupture, unlike other magnitude scales that depend upon the amplitude of the seismic waves. Larger earthquakes have larger seismic moment (product of rupture area and average fault displacement). The moment magnitude scale, therefore, does not saturate with the size of the earthquake and also provides the ability to distinguish between a large and a great earthquake. For example, the surface wave magnitude for both the 1906 San Francisco Earthquake and the 1960 Chile Earthquake was estimated as 8.3, although the rupture area of the 1960 earthquake was about 35 times greater than that of the 1906 earthquake. The moment magnitude for the two earthquakes has been computed to be approximately 8.0 and 9.5, respectively (Reiter 1990).

Figure 3.A-1 shows an approximate relation between the various magnitude scales. According to this figure, except for M_S, all other magnitude scales are similar to M_W up to the point where they start to saturate.

Another descriptor of the size of an earthquake is the intensity at a given site. The intensity at any location is described by a qualitative scale that uses eyewitness accounts of fault motion and damage assessments to describe the amount of movement felt at that location. One commonly used scale is the Modified Mercalli intensity (MMI). As Table 3.A-1 shows, the MMI scale consists of 12 intensity levels. Whereas intensity I means the earthquake is practically not felt, intensity XII indicates almost total destruction. Intensity XII is rare except in very large earthquakes. Intensity X can occur in moderate to large earthquakes, especially in areas close to the rupture zone. Engineered structures can be damaged in areas experiencing intensities in the range of VIII–X.

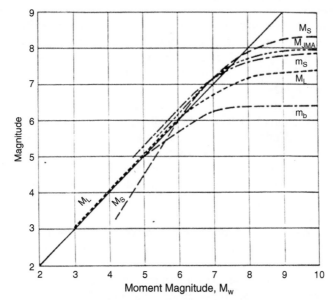

Figure 3.A-1. Relation between Moment Magnitude and other magnitude scales: M_L (local), M_s (surface wave), m_b (short-period body wave), m_g (long-period body wave), and M_{JMA}.
Source: Japan Meteorological Agency.

3.A.4 SITE-SPECIFIC RESPONSE SPECTRA

Two distinct approaches are generally used for site-specific estimates for developing seismic design criteria for engineered projects: the deterministic procedure and the probabilistic procedure. Descriptively, the two procedures would at first appear to be quite different with a deterministic approach offering the advantage of appearing easiest to follow. Properly applied, either procedure can lead to satisfactory seismic design.

In areas of low seismic hazard, the additional design and material costs that ensue when unsure but conservative decisions are made may not be excessive. When the seismic exposure is high, the cost associated with additional conservatism can increase significantly. This increase needs to be balanced with the likelihood that the damaging event may or may not occur during the life of the structure. Large damaging earthquakes are infrequent events that may simultaneously affect a large number of structures. The extent of damage during an earthquake can be related to the level of motions generated by the earthquake. The level of earthquake motion against which design or evaluation criteria should be developed requires subjective judgment based on experience and observation. Probabilistic methods were developed as a result of a desire to quantify some of

Table 3.A-1. Modified Mercalli Intensity Scale.

I. Not felt. Marginal and long-period effects of large earthquakes.

II. Felt by persons at rest, on upper floors, or favorably placed.

III. Felt indoors. Hanging objects swing. Vibration like passing of light trucks. Duration estimated. May not be recognized as an earthquake.

IV. Hanging objects swing. Vibration like passing of heavy trucks, or sensation of a jolt like a ball striking the walls. Standing motor cars rock. Windows, dishes, doors rattle. Glasses clink. Crockery clashes. In the upper range of IV wooden walls and frames creak.

V. Felt outdoors; direction estimated. Sleepers wakened. Liquids disturbed, some spilled. Small unstable objects displaced or upset. Doors swing, close, open. Shutters, pictures move. Pendulum clocks stop, start, change rate.

VI. Felt by all. Many frightened and run outdoors. Persons walk unsteadily. Windows, dishes, glassware broken. Knickknacks, books, etc., off shelves. Pictures off walls. Furniture moved or overturned. Weak plaster and masonry D cracked. Small bells ring (church, school). Trees, bushes shaken (visible, or heard to rustle).

VII. Difficult to stand. Noticed by drivers of motor cars. Hanging objects quiver. Furniture broken. Damage to masonry D, including cracks. Weak chimneys broken at roof line. Fall of plaster, loose bricks, stones, tiles, cornices (also unbraced parapets and architectural ornaments). Some cracks in masonry C. Waves on ponds; water turbid with mud. Small slides and caving in along sand or gravel banks. Large bells ring. Concrete irrigation ditches damaged.

VIII. Steering of motor cars affected. Damage to masonry C; partial collapse. Some damage to masonry B; none to masonry A. Fall of stucco and some masonry walls. Twisting, fall of chimneys, factory stacks, monuments, towers, elevated tanks. Frame houses moved on foundations if not bolted down; loose panel walls thrown out. Decayed piling broken off. Branches broken from trees. Changes in flow or temperature of springs and wells. Cracks in wet ground and on steep slopes.

IX. General panic. Masonry D destroyed; masonry B seriously damaged. (General damage to foundations.) Frame structures, if not bolted, shifted off foundations. Frames racked. Serious damage to reservoirs. Underground pipes broken. Conspicuous cracks in ground. In alluviated areas, sand and mud ejected, earthquake fountains, sand craters.

(Continued)

Table 3A-1. Modified Mercalli Intensity Scale. (Continued)

X.	Most masonry and frame structures destroyed with their foundations. Some well-built wooden structures and bridges destroyed. Serious damage to dams, dikes, embankments. Large landslides. Water thrown on banks to canals, rivers, lakes, etc. Sand and mud shifted horizontally on beaches and flat land. Rails bent slightly.
XI.	Rails bent greatly. Underground pipelines completely out of service.
XII.	Damage nearly total. Large rock masses displaced. Lines of sight and level distorted. Objects thrown into the air.

Notes: To avoid ambiguity, the quality of masonry, brick, or other material is specified by the following lettering system. (This has no connection with the conventional classes A, B, and C construction.)

Masonry A. Good workmanship, mortar, and design; reinforced, especially laterally, and bound together by using steel, concrete, etc.; designed to resist lateral forces.

Masonry B. Good workmanship and mortar; reinforced, but not designed to resist lateral forces.

Masonry C. Ordinary workmanship and mortar; no extreme weaknesses, like failing to tie in at corners, but neither reinforced nor designed to resist horizontal forces.

Masonry D. Weak materials, such as adobe; poor mortar; low standards of workmanship; weak horizontally.

this judgment understanding and to allow other practitioners to use it in a repeatable way.

The following sections discuss both the deterministic and probabilistic seismic hazard approaches.

Deterministic Seismic Hazard Analysis

In a deterministic approach, ground motions at a site are estimated by considering a single event of a specified magnitude and distance from the site. To perform a deterministic analysis, the following data are required:

(a) Definition of an earthquake source (e.g., a known geologic fault) and its location relative to the site;

(b) Definition of a design earthquake that the source is capable of producing; and

(c) A relationship that describes the attenuation of the ground motion parameter of interest, for example, peak ground acceleration or response spectral ordinates for a specific natural period or frequency of vibration.

An earthquake source in most cases is a known active or potentially active geologic fault. A site may have several faults nearby. All these sources must be identified. Based on the length and characteristics of the fault, a maximum magnitude potential of each source is specified. Estimation of maximum magnitude potential of geologic sources can be obtained from published sources or qualified professionals. The maximum magnitude potential of a source is determined from empirical relationships between magnitude and rupture length or through historical knowledge of past earthquakes on the particular source. Geologists estimate the maximum magnitude potential based on the character-istics of the source and factors such as historical activity and source dimensions. Several relationships exist that relate source dimensions to its maximum magni-tude potential. This information is then used as input to an attenuation relation-ship, which relates the ground motion parameter of interest to magnitude and distance.

Attenuation of ground motion is known to vary significantly during any earthquake. While the variation is widely known, the quantification of the reasons for the variation is not fully understood. Examples of possible causes may be different rupture mechanisms in the epicenter area, directionality of the radiation pattern of the motion, and regional geological variations. Researchers using statistical analysis of large ground motion data sets have shown that the scatter of ground motion values about a mean attenuation relationship can be represented by a log-normal distribution. Attenuation relationships are empirical relationships developed by using regression analysis techniques on ground motion data recorded during earthquakes. Several different attenuation relationships are available in published sources and are a function of earthquake magnitude, its distance, local geology, and tectonic environment (subduction zones of the Pacific Northwest, eastern United States, and western United States). Power et al. (2008) summarize the recent attenuation relationships.

Because attenuation relationships are empirical relationships derived using statistical analysis techniques on recorded data, they do not fit each and every data point exactly. The actual recorded data spreads around the median attenuation relationship. This spread is defined by the standard deviation. The attenuation equation with no standard deviation means that 50% of the recorded values are above and 50% are below the predicted value. An attenuation relationship with one standard deviation means that 84% of the recorded values are below the predicted values; similarly, attenuation relationships with two standard deviations imply that 98% of the recorded data are below the predicted value.

The definitions of earthquake magnitude and source-to-site distance some-times vary from one attenuation relationship to another. The magnitude scale and source-to-site distance used in the analysis must be consistent with the proper definition for the particular attenuation relationship used. If the site has several sources in close proximity, an envelope of values of response spectra developed for each source can be used.

Probabilistic Seismic Hazard Analysis

Probabilistic seismic hazard analyses (PSHA) require additional seismic criteria beyond the selected magnitude, minimum distance, and the attenuation equations used for the deterministic procedure defined in the previous section. The primary additional requirements are the dimensions of each source region and the means to interpret the probability of occurrence within each source.

Probabilistic seismic hazard analysis procedures were first developed and used by Cornell (1968) and Cornell and Vanmarcke (1969) and further extended by Merz and Cornell (1973) and Der Kiureghian and Ang (1977). This method involves careful consideration and incorporation of the geology, history, and tectonics of the site region into a seismotectonic source model. A probabilistic seismic hazard study requires the acceptance of several assumptions regarding earthquakes.

The first step in PSHA requires that the seismic history and geology are sufficiently well known to permit an estimate of the region's seismic activity. This is defined as the region's seismicity. This information is then used to develop a seismic source model. Potential sources can be defined as either point sources, line sources, or area sources (Der Kiureghian and Ang 1977). Considerations may be needed to account for the random occurrence of small or moderate sized earthquakes (background seismicity).

The probability distribution of earthquakes to obtain the relative occurrence frequency distribution between large and small earthquakes is then obtained. Estimation of recurrence rate is a key parameter in PSHA procedures. Two approaches for estimating the recurrence rate are generally used. One is based on the historical seismicity records, and the other is a geologic slip rate model. The former is based on the recorded seismic history of the region, whereas the latter uses geologic information to define strain rate accumulation using long-term slip-rate data for a fault.

Recurrence rates can be expressed in the form of the familiar Richter's law of magnitudes (Richter 1958), expressed as

$$\log N = a - bM \qquad (3.A\text{-}4)$$

where N is the number of events of magnitude M or greater for the time period under consideration, and a and b are constants that depend on the seismicity of the region.

Other recurrence models, such as the characteristic earthquake model (Youngs and Coppersmith 1985), may also be used.

For PSHA, the minimum value of 5.0 is usually considered for M, because earthquakes of sizes less than 5.0 are generally not damaging to engineered structures. Due to the integrative nature of PSHA, including earthquakes of magnitude less than 5.0 would tend to increase the ground motion hazard for the peak ground acceleration and short periods of the response spectrum for small return periods because of the relatively high frequency of occurrence of small earthquakes, which are not necessarily damaging to structures.

The PSHA procedures combine the probability of occurrence of an earthquake, the probability of it being a specific size, and the attenuation of its motion to the site, to obtain the probability of exceedance of a specified ground motion level. Combination of these individual probabilities for the different source zones at each motion level of interest then provides the total annual probabilities of exceedance. Ground motions corresponding to a particular probability level can then be obtained by interpolation. Figure 3-4 of Chapter 3 shows the different elements of a PSHA graphically. Several computer programs are available for PSHA.

The choice of probability levels for ground motion usually depends upon the design criteria for each individual project and is established based on several parameters such as level of risk and importance of the facility. Various probability levels commonly used are 50% probability of exceedance in 50 years, 10% probability in 50 years, and 10% probability of exceedance in 100 years. Assuming a Poisson distribution, these probability levels correspond to an equivalent return period of 72-, 475-, and 950-year return periods, respectively, or annual exceedance probabilities of 0.014, 0.002, and 0.001. The relation between return period, T, and probability of exceedance for a Poisson distribution is given as

$$T = \frac{-t}{\ln(1 - p)} \tag{3.A-5}$$

where p is the probability of exceedance in t years. In this equation, the probability of exceedance, p, is expressed as a decimal (e.g., 0.10 stands for 10%).

A response spectrum developed from a PSHA is referred to as a uniform hazard spectrum, because each ordinate has the same associated probability of exceedance (a constant level of risk). Because the response spectrum from a PSHA analysis is calculated by a sum of individual probabilities of occurrence of several earthquakes of different magnitudes and distances within the entire area of study, it cannot be related to an earthquake of specified magnitude and distance from the source.

The level of ground motion corresponding to a 2% probability of exceedance in 50 years is the basis of the maximum considered earthquake (MCE) in ASCE 7. In the Unified Building Code (UBC 1997), the design basis is the ground motion with a 10% probability of exceedance in 50 years.

Uncertainty associated with the selection of input parameters must be included in a PSHA. Uncertainty could be in the definition of magnitude-recurrence relationship, or in the definition of slip rates, or the maximum magnitude potential, or geographical location of a source. Furthermore, uncertainty also exists in defining the ground motions. The uncertainty distribution on the attenuation model indicates that no matter how accurately we know the magnitude and distance of a postulated earthquake, some uncertainty will still remain in predicting what the ground motion will be (Reiter 1990). The uncertainty associated with an attenuation relationship (standard deviation) is incorporated in the seismic hazard during the integration process to calculate the probability of occurrence of a specified ground motion level.

Sometimes a tendency exists to make conservative estimates of input parameters to be used as an input to a PSHA. This is not consistent with the philosophy of PSHA. Input parameters should be selected based on "best estimate" values. The uncertainty on the best estimate can be incorporated in the analysis using a logic tree approach, where multiple estimates are specified and weighted.

APPENDIX 3.B EARTHQUAKE-RELATED COASTAL INUNDATION

3.B.1 INTRODUCTION

Section 3.6 describes briefly the phenomena associated with tsunamis and seiches. This appendix discusses in more detail several issues associated with earthquake-induced coastal inundation.

3.B.2 EARTHQUAKE-INDUCED COASTAL INUNDATION HAZARDS

One or more of the following earthquake-related inundation hazards may affect a given site in a coastal area:

(a) *Coastal subsidence* caused by tectonic (faulting) or nontectonic (e.g., submarine landslide) effects;

(b) *Tsunamis* from either large ($M_W > 6.5$) local earthquakes or distant great ($M_W > 7.7$) earthquakes; and

(c) *Seiches* in semi-enclosed bays, estuaries, lakes, or reservoirs caused by moderate ($M_W > 5.0$) local (submarine) earthquakes or regional large earthquakes.

Each of these hazards is described in detail below.

3.B.2.1 Coastal Subsidence

In areas of dip-slip faulting (normal, thrust, or oblique-slip faulting) that involves a vertical component of movement, coseismic subsidence along coastal faults could result in permanent submergence of the coastal area. Based on global experiences such as the 1964 Alaska ($M_W = 8.4$); the 1992 Cape Mendocino, California ($M_W = 7.2$); and the 2004 Banda Aceh, Indonesia earthquakes, tectonic subsidence of more than 6 ft (2 m) could occur. Furthermore, earthquakes on blind faults, such as the 1994 Northridge California Earthquake ($M_W = 6.7$), may create significant surface uplift or subsidence without attendant surface fault rupture.

Permanent coastal subsidence may also occur as a result of earthquake-induced seafloor slumping or landslides. During the 1964 Alaska Earthquake, the Seward port area, which included petrochemical facilities, progressively slid under the bay as shaking continued. This subsidence was followed by withdrawal of the sea and subsequent inundation of the coast by a wave 30 ft high. In addition to this slide-induced wave, tsunami waves some 30 to 40 ft high, generated by the tectonic deformation, arrived later and compounded the destruction of Seward. Similar effects on a lesser scale were experienced at Moss Landing on Monterey Bay during the 1989 Loma Prieta, California Earthquake ($M_W = 7.1$) even though the epicenter was located onshore in the Santa Cruz Mountains. At Moss Landing, subsidence and related liquefaction created large fissures, generated a small tsunami in Monterey Bay, and caused substantial damage to the Moss Landing Marine Facilities. Thus, sites located where landslides or unstable slopes are common, both on land and offshore, should assume that submarine slumping and coastal inundation could occur during large earthquakes in the region.

For disaster planning and emergency response purposes, significant coastal subsidence followed by tsunami run-up should be considered likely to exceed run-up predicted based on presubsidence topography alone. Combined subsidence and tsunami effects are considered the likely cause of the destruction to the city of Port Royal on the Island of Jamaica in 1692 and could occur in many seismically active areas of the world. Thus, maps of potential inundation should consider these combined effects where coastal subsidence may be expected.

3.B.2.2 Tsunami (Seismic Sea Wave)

Tsunamis are long-period (T from 5 to 60 min) surface gravity waves, with wavelengths that may exceed 60 mi (100 km). Tsunamis are typically generated by large submarine earthquakes that displace the seafloor over large areas. The destructive effects of tsunami waves may be localized, occurring along the coasts situated close to the tsunami origin, or the waves may travel with little attenuation across entire oceans and affect coasts thousands of miles away.

Earthquake-induced seafloor displacement may occur with or without sea-floor fault rupture (as on a blind fault) and could generate a potentially destructive tsunami. Even strike-slip faults, although less likely to generate significant vertical seafloor displacement than dip-slip faults, may cause substantial seafloor uplift or subsidence in places where these faults bend and curve. Furthermore, large earthquakes centered on faults near the coast but on land (e.g., the 1992 Cape Mendocino, California Earthquake) may also cause local tsunamis because of the broad regional tectonic uplift (or subsidence) they create.

In addition to primary tectonic deformation associated with the earthquake, large-scale slumping or landslides under the ocean or other large water bodies may generate significant local tsunamis. A large rockslide in Lituya Bay, Alaska, triggered by an earthquake in 1957 generated a tsunami that surged 1,700 m up the side of the fjord. Volcanic explosions, such as that of Krakatoa in 1883, may also generate destructive tsunamis that inundate surrounding coasts. Coastal areas in active tectonic regions are more likely to be susceptible to tsunami attack. However, large tsunamis may travel across entire ocean basins and affect areas with little or no local tectonic activity.

Around the Pacific Ocean basin, often referred to as the "Ring of Fire," most of the world's great earthquakes, volcanic eruptions, and destructive tsunamis occur. Earthquakes in this region have large dip-slip (vertical) movements that displace the seafloor rapidly, initiating the sea surface disturbance that becomes a tsunami. Places such as Japan, Alaska, Hawaii, and western South America are particularly vulnerable to these tsunamis, both because they have many large earthquakes (and volcanoes) locally situated and because their geographic position and coastal configuration expose them to distantly generated tsunamis. In the contiguous United States, recent evidence uncovered along the coasts of Washington, Oregon, and northern California show that large local tsunamis have been generated by great earthquakes in the Cascadia subduction zone. The Atlantic coast has also suffered destructive tsunamis such as the one generated by the great Lisbon Earthquake in 1755, but such events are rare and generally less severe on the US side of the Atlantic Ocean. Whereas the Gulf of Mexico has been relatively free of destructive tsunamis, in the Caribbean, active tectonics along the Puerto Rico trench and Antilles arc have generated notable tsunamis. Active tectonism with large earthquakes, volcanoes, and tsunamis also occurs around the Indian Ocean.

Propagation of tsunamis often results in dispersion of the initial tsunami wave form so that several significant waves may strike the coast. The first wave to arrive may be smaller than subsequent waves, and persons in affected areas must realize that a drawdown following one wave may soon be followed by larger, more destructive, wave arrivals. The run-up speed of such waves (1 to 20 m/s or 20 to 40 mi/h) would outpace even the fastest human runners. In bays or other partially enclosed basins, tsunamis may also set up a seiche (see Section 3.B.2.3), which can amplify wave height and prolong the tsunami's duration. For example, following the 1964 Alaska Earthquake ($M_S = 8.4$), the tsunami recorded at Santa Monica, California, measured up to 2 m (6.5 ft) high and oscillations of half this amount continued to occur 17 to19 hours after the arrival of the first wave. Consequently, destructive waves from a tsunami may persist for a long time, and vulnerable areas should remain evacuated until authorities have broadcasted an "All Clear" message.

3.B.2.3 Seiche

A seiche occurs when resonant wave oscillations form in an enclosed or semi-enclosed body of water such as a lake, bay, or fjord. Seiches may be triggered by moderate or larger local submarine earthquakes, and sometimes by large, distant earthquakes. Tectonic deformation of Hebgen Lake during the 1959 earthquake initiated seiche oscillations of up to 8 ft over the new lake level (which changed as a result of the deformation). Such oscillations have also been triggered by meteo-rological disturbances. For example, the passage of a storm front across Lake Michigan tends to pile water on the eastern shore as the front advances. After the front passes, the water flows back westward, setting up the seiche. The initial storm wave setup is similar to storm surges observed along coasts affected by hurricanes and other tropical storms, but the seiche results from the resonant oscillation of the wave due to the enclosed character of the water body. Seiche oscillations may persist for several hours.

3.B.3 SITE VULNERABILITY ASSESSMENT

Coastal sites in active tectonic areas where dip-slip faulting is common are most susceptible to earthquake-induced inundation. Examination of the regional fault characteristics and earthquake history, with emphasis on submarine earthquakes and possible historical tsunamis, provides the best measure of site vulnerability. Estimates of maximum tsunami run-up may be based on either historical occurrences or theoretical modeling. The Federal Insurance Administration developed inundation maps to assist in setting rates for the National Flood Insurance Program (flood insurance rate maps or FIRM). Inundation levels on these maps generally consist of the higher of riverine flood, coastal storm surge, or tsunami inundation levels. FIRM show coastal run-up elevations for 100- and 500-year events based upon theoretical tsunami modeling for the US Pacific Coast (Houston and Garcia 1978). Some cities, particularly in Hawaii where tsunamis are relatively frequent, have prepared local tsunami inundation maps. These should be available from government agencies. Eastern US cities (Atlantic and Gulf coasts) are more likely to be affected by hurricane storm surge inundation, whereas Pacific Coast cities are to be affected by tsunamis. Inland cities are affected by riverine flooding, although cities adjacent to major lakes, such as Chicago, are also susceptible to seiche. If a site lies below the inundation elevations, then it may be vulnerable to flooding and related effects. Lower elevations are also more vulnerable because deeper water levels carry greater hydrodynamic forces for impact and erosion of structures and foundations. Table 3.B-1 outlines the steps involved in assessing the potential for coastal inundation.

3.B.4 EFFECTS OF COASTAL INUNDATION

Any of the three causes of coastal inundation (coastal subsidence, tsunami, or seiche) results in essentially the same effect: low-lying coastal areas will be covered with water. The greatest hazard from coastal inundation results from the inflow and outflow of the sea during the inundation event. The strong currents from this flow may erode foundations of structures and sweep away smaller structures or equipment. Debris carried by the water will act as battering rams to pound other more solidly anchored structures. People unsafely located in the low-lying areas may be swept out to sea and drowned or knocked unconscious by the debris and drowned. Normal wave action will also be superimposed on the tsunami, adding dynamic forces capable of pounding and destroying even strong structures. Petrochemical facilities would be especially vulnerable to rupture of storage tanks from debris impact and foundation erosion, possibly resulting in explosion and fire. The tsunami at Seward, Alaska, following the 1964 earthquake led to such destruction of the port facilities. The succession of waves spread the fire from ruptured oil tanks across and throughout the port area.

Tsunami combined with seiche could result in an oscillating water level causing several cycles of coastal inundation. Periods of inundation could last for several hours. With tectonic subsidence of the coast (or from a large submarine landslide), the inundation may be permanent. Major subsidence, ground

Table 3.B-1. Evaluation Process for Coastal Inundation Hazards.

Outline for Analysis of Coastal Inundation Potential
(1) Evaluate tsunami and earthquake history
(2) Inundation maps prepared by government agencies (e.g., FIRM, Houston and Garcia 1978)
(3) Seismotectonic study, character of faulting, location of faulting with respect to the site
(4) Site topography and adjacent coastal bathymetry, harbor or embayment configuration (resonances)
(5) Hydrodynamic analyses, long-period wave response, tidal parameters

Tsunami/Seiche Inundation Evaluation Procedure
A. Determine whether tsunami/seiche/inundation is possible (history, seismotectonic study):
  1. Literature review
  2. Historical research
  3. Seismotectonic analysis
  4. Geomorphic/paleoseismic study
  5. Geotechnical study (landslides, slumps, possible generating mechanisms)
B. Determine maximum credible run-up elevation for region:
  1. Tsunami inundation maps available?
  2. FIRM
  3. Historical analysis
  4. Theoretical analysis
C. Determine whether site is located within low-lying coastal area (below maximum possible run-up elevation)
D. Engineering analysis required to evaluate tsunami/seiche inundation potential
E. Engineering analysis required to identify vulnerable facility components
F. Engineering analysis required to identify/recommend specific mitigations for vulnerable facility components
  1. Seawalls or barriers
  2. Facility layout to locate vulnerable components above inundation zone
  3. Geotechnical studies to identify potential liquefaction/slump/landslide hazards that might place facility at risk (Seward, Alaska, problems)
  4. Facility construction with reinforced foundations and walls below inundation levels to resist tsunami damage
  5. Response plans: tsunami warning system, evacuation to high ground, safe shutdown, automatic fire-fighting equipment(?), life-saving (flotation) devices
G. Government studies available/needed for region?

deformation, or severe erosion could undermine bridge piers and foundations, disrupting lifelines such as highways, railroads, pipelines, and electrical power and communication lines. Such damage could isolate the area from emergency equipment, supplies, and rescue.

3.B.5 MITIGATION

3.B.5.1 Shoreline Structures

Throughout Japan, along the coast where tsunami attack is common, sea walls and other barriers have been constructed to resist tsunami inundation. These structures are sometimes unsuccessful, as large tsunamis often overtop the barriers and flood the protected area, generating strong currents that erode the barriers and other structures. In some instances, large wave dissipation structures have been carried hundreds of meters inland by major tsunamis. These structures could act as powerful battering rams against the very structures they were designed to protect. Even large storm surf sometimes throws large boulders and rip-rap inland causing damage to structures. Therefore, coastal protective structures must be carefully designed to withstand extreme events.

3.B.5.2 Tsunami Warning

Tsunamis travel across the oceans at high speeds of more than 60 mi/h (100 km/h), but generally require several hours before reaching distant coasts. Therefore, tsunami warnings are possible and have been established through the Pacific tsunami warning network. Citizens must take these warnings seriously: Failure to heed the warnings by going to the beach to watch the incoming waves may result in needless loss of life. In southern California, the great hazard that tsunamis pose is seriously underappreciated. This perhaps results from the absence of significant tsunamis striking the Southern California coast in the recent past (one or two generations). Southern California beaches, such as Santa Monica, are most vulnerable to distant tsunamis arriving from the south or north. Such tsunamis would be generated by large or great earthquakes along the South American or Alaska and Aleutian coasts. Other coastal areas would be vulnerable to tsunamis from sources located along direct paths across the ocean. For large, approximately linear tsunami sources such as oceanic trench areas, directional effects tend to focus the tsunami energy along an axis perpendicular to the linear source region. Sites in Hawaii are vulnerable to attack from tsunamis in many directions because of its central location in the Pacific Ocean.

A minimum of about 20 to 30 min is required to issue a tsunami warning through the Pacific tsunami warning network. Locally generated tsunamis would arrive too quickly, in less than 10 to 15 min, for adequate warning by officials. Although the technology exists to develop more rapid, regional tsunami warning networks, the relatively rare occurrence of damaging local tsunamis and economic constraints have precluded installation of such a system in most areas (such as the Southern California area). In the absence of official warning systems, persons in low-lying coastal areas vulnerable to tsunami inundation should seek shelter on higher ground immediately upon experiencing a strong earthquake in their area. Such immediate response has been conditioned into the minds of Japanese people where tsunami occurrences are relatively frequent. If no tsunamis were generated in the first hour following the event, and no tsunami warnings were issued based upon other earthquake data, returning to the low-lying area should be safe. If, however, a tsunami is generated, the first wave may arrive and withdraw causing little or no damage, only to be followed by larger, possibly destructive waves. For areas that tend to have resonant oscillations, significant wave activity may persist for several hours, or even most of a day. People in these areas should wait for an "All Clear" signal from official agencies before returning.

3.B.5.3 Land-Use Planning and Inundation Map

Theoretical and laboratory models have been used to estimate likely inundation levels for 100- and 500-year return periods along US coasts affected by tsunami (FIRM). The 100- and 500-year time intervals represent an estimated average recurrence interval; however, so-called 100-year events may occur in shorter time spans. For example, both the 1960 Chile and 1964 Alaska earthquakes generated tsunami waves equivalent to a 100-year event along the Southern California coast. The run-up height estimated in these maps refers to the land elevation (or contour) that the incoming tsunami will reach. This run-up height includes the effects of the normal tidal range, but excludes the possible simultaneous occurrence of the tsunami with high surf and storm wave setup. Amplification of run-up by coastal topography may also occur, especially in narrow, V-shaped canyons.

Sites located in vulnerable areas should prepare for the possibility of tsunami inundation, and important structures should be designed to resist possible inundation effects. Facilities should be evaluated to identify vulnerable components, and possible mitigation strategies identified. Emergency or disaster response plans should be prepared considering the vulnerability of the facility and the possible effects of tsunami attack. Also, an attempt should be made to anticipate otherwise unexpected consequences that may arise from the complex interaction between direct earthquake effects and subsequent tsunami inundation.

3.B.5.4 Public Awareness

McCulloch (1985) lists the following individual actions that can help to save lives in the event of a tsunami:

(a) If you are on low ground near the coast and a large earthquake occurs in your area, move to high ground. There may be no time to either issue or receive an official warning.

(b) If you recognize a drawdown of the sea, move to high ground.

(c) The first tsunami wave may not be the highest. In major tsunamis, later waves commonly have the highest run-up heights.

(d) Periods between successive major waves may be similar. Thus, you may have time between waves to move to higher ground or to assist in rescue efforts.

(e) Do not assume that because the incoming tsunami wave is not breaking it will not be destructive. The forces contained in this high-velocity, often debris-laden torrent are extremely destructive during run-up and run-off.

(f) Do not assume that areas behind beaches generally shielded from storm waves will be immune from high run-up. Tsunami run-up heights have historically been higher in some such areas along the California coast.

(g) Do not go to the beach to watch a tsunami coming in. Not only might you hamper rescue efforts, but you may also have taken your last sightseeing trip.

References

Houston, J. R., and A. W. Garcia 1978. *Type 16 flood insurance study: Tsunami prediction for the West Coast of the Continental United States.* Technical Rep. No. H-78. Vicksburg, MS: US Army Engineer Waterways Experiment Station.

McCulloch, D.S. 1985. *Evaluating tsunami potential, in evaluating earthquake hazards in the Los Angeles region: An earth science perspective: USGS professional paper 1360,* 375–413. Reston, VA: USGS.

CHAPTER 4

Seismic Analysis

4.1 INTRODUCTION

This chapter describes the overall methodology for performing seismic analysis of petrochemical facilities to obtain seismically induced loads. This is applicable to structural systems or subsystems to be used either in new design or in evaluation of existing facilities. This chapter is organized such that the main contents are written primarily for design of new facilities. Section 4.6 summarizes considerations for the evaluation of existing facilities.

The overall approach outlined in this chapter is consistent with the specific requirements for nonbuilding structures found in ASCE 7-16. ASCE 7 is the basis for the seismic load provisions of the IBC (2018).

The intent of this chapter is not to instruct an engineer on how to use ASCE 7 or any other commonly used building code. Rather, it is to give guidance as to appropriate application of typical building code provisions to the types of structures and equipment commonly found in petrochemical facilities. The reader should be aware that building code provisions are constantly being revised.

Users of codes other than ASCE 7 are encouraged to read those sections of this chapter that primarily use ASCE 7 equations as a reference, such as the discussion on the base shear equation in Section 4.4.2. Much of the intent behind the guidance for application to petrochemical structures will still be applicable to other codes. In previous editions of this *Guideline*, specific guidance was given in many cases that differed significantly from the ASCE 7 provisions. In all cases, this specific guidance was incorporated into the current ASCE 7 or IBC provisions.

Seismic building codes used in the past emphasized building structures and were developed with a primary focus on methods and earthquake experience from building performance. ASCE 7 incorporates seismic provisions developed through the National Earthquake Hazards Reduction Program (NEHRP) process specifically for use with nonbuilding structures. This chapter emphasizes application to nonbuilding structures not similar to buildings and gives guidance as to the use of the seismic provisions of ASCE 7. The appendixes also give examples of application to typical petrochemical installations, such as vessels, elevated equipment, process columns, pipeways, and so on. Much of the guidance in the appendixes, such as calculation of structural periods in Appendix 4.A, is directly applicable to users of any building code.

In addition, Section 4.6 and several appendixes of this chapter discuss considerations for assessing existing structures. (All building codes primarily focus on design of new facilities.) Appendixes 4.D and 4.E provide alternate methods for evaluating the overturning potential and the potential for sliding commensurate with the overall philosophy for the evaluation of existing structures. Appendix 4.F presents a guidance document prepared by others that is used in many locations in California to assist in evaluating existing facilities in relation to the requirements of a state-mandated regulatory program. This document is presented in its entirety in Appendix 4.F as an aid to interested engineers.

Some of the material presented in this chapter, including appendixes, was taken from various companies (architect-engineer, consultant, and owner) currently active in the petrochemical industry. Appropriate references are provided to guide the reader to the background of the presented material.

4.2 STRUCTURAL SYSTEMS IN A PETROCHEMICAL FACILITY

4.2.1 General

Structures in petrochemical facilities fall into two main categories: building structures (Section 4.2.2) and nonbuilding structures (Section 4.2.3). Within nonbuilding structures are two subcategories defined depending on structural characteristics. These are nonbuilding structures similar to buildings (Section 4.2.3.1) and nonbuilding structures not similar to buildings (Section 4.2.3.2). Section 4.2.3.2 also addressed combination structures. In addition, certain equipment and anchorage design requirements in a petrochemical facility fall under the category of "subsystem" rather than structure. Section 4.2.4 discusses nonstructural components and systems. Esfandiari and Summers (1994) briefly describe various structures typically found in petrochemical facilities.

4.2.2 Building Structures

Building structures typically found in petrochemical facilities include administration buildings, control buildings, substations, warehouses, firehouses, maintenance buildings, and compressor shelters or buildings. They are typically single-story buildings, but may have as many as two or three stories. Seismic force–resisting systems (SFRSs) used include shear walls, braced frames, rigid frames, and combinations thereof. The provisions of the code of record would apply in their entirety to design of these buildings. Certain buildings in petrochemical facilities may also be required to be designed for blast loading, which often controls the design.

4.2.3 Nonbuilding Structures

Other than actual buildings in a petrochemical facility, all structures are typically classified as nonbuilding structures. However, their structural systems may resemble those of buildings, for example, transverse moment frames for pipeways. Therefore, these structures are classified as nonbuilding structures similar to buildings. Other structures, whose structural systems do not resemble those of buildings, are classified as nonbuilding structures not similar to buildings. An example is a tank or a vessel. The following sections define these subcategories in detail.

4.2.3.1 Nonbuilding Structures Similar to Buildings

A nonbuilding structure similar to a building is a structure that is designed and constructed in a manner similar to buildings and that will respond to strong ground motion in a fashion similar to buildings. These are structures such as pipeways, equipment support frames, and box-type heaters that have SFRSs similar to those of building systems, such as braced frames, moment-resisting frames, or shear wall systems. A flexible structure is typically defined as having a natural period of vibration (T) of 0.06 s or more, which is equivalent to a frequency of about 17 Hz or less. Examples of building-like structures found in petrochemical facilities include

(a) Moment-resisting frames (steel or concrete) or braced frames (cross-braced or chevron-braced) supporting exchangers and horizontal vessels. Such structures can be up to four or five levels high.

(b) Pipeways with SFRSs that are moment-resisting frames (usually in the transverse direction to provide access beneath the pipeway) or braced frames (usually in the longitudinal direction).

(c) Rectangular furnaces.

Many building code requirements are keyed to the occupancy of the structure. Because buildings and nonbuilding structures similar to buildings use many of the same structural systems and physically resemble each other, building officials often incorrectly classify nonbuilding structures similar to buildings as buildings. Nonbuilding structures similar to buildings are typically structures that plant personnel visit infrequently to perform brief tasks or to monitor a process. Most building officials deal with the term "occupant load," which is measured in terms of the number of persons for which a building's means of egress is designed. Occupant load is best compared with peak occupancy load as found in API RP 752 (2009). Designating a nonbuilding structure similar to a building as "unoccupied" or determining the peak occupancy of the structure can help to prevent the incorrect classification of a structure. ASCE 7-16, Section 11.1.3, also clarifies that "Buildings whose purpose is to enclose equipment or machinery and whose occupants are engaged in maintenance or monitoring of that equipment, machinery, or their associated processes shall be permitted to be classified as nonbuilding structures designed and detailed in accordance with Section 15.5 of this standard."

4.2.3.2 Nonbuilding Structures not Similar to Buildings

Nonbuilding structures not similar to buildings are structures that lack lateral and vertical seismic force–resisting systems like those of buildings. This category covers many structures and self-supporting equipment items found in a typical petrochemical facility, such as vertical vessels, horizontal vessels and exchangers, stacks, and towers. Nonbuilding structures not similar to buildings typically found in a petrochemical facility fall into one of four categories described as follows:

(a) Rigid structures, that is, those whose fundamental structural period is less than 0.06 s, such as a horizontal vessel or exchanger, supported on short, stiff piers.

(b) Flat bottom tanks supported at or below grade: Such structures respond very differently (compared with regular structures) during an earthquake. Special issues for unanchored tanks, such as the effects of fluid sloshing and tank uplift, must also be considered. Chapter 7 describes tanks in more detail.

(c) Other nonbuilding structures not similar to buildings: Examples of this category of structure include skirt-supported vertical vessels, spheres on braced legs, horizontal vessels or exchangers on long piers, guyed structures, and cooling towers.

(d) Combination structures: In petrochemical facilities, such structures generally support flexible nonstructural elements whose combined weight exceeds about 25% of the weight of the structure. A typical example is a tall vertical vessel, furnace, or tank supported above grade on a braced or moment-resisting frame. The analysis method depends on whether the nonstructural element is flexible or rigid and whether its weight exceeds or is less than about 25% of the supporting structure's weight. Appendix 4.B describes these analysis methodologies.

4.2.4 Nonstructural Components and Systems

In addition to various structures described in the preceding sections, various nonstructural components and systems are supported within a structure. The weight of nonstructural components and systems should only be a small portion of the combined weight of the supporting structure and nonstructural component (i.e., less than about 25%). Examples of nonstructural components and systems typically found in a petrochemical facility include

(a) Horizontal vessels and exchangers supported on a structure, weighing less than about 25% of the combined weight of the supporting structure and nonstructural component;

(b) Electrical and mechanical equipment supported within a structure;

(c) Cable tray, conduit, ductwork, and/or piping supported on a pipeway or within a building; and

(d) Suspended ceilings, cabinets, and lighting fixtures.

4.3 SELECTION OF ANALYSIS PROCEDURES

4.3.1 General

When performing seismic analysis of structures in petrochemical facilities, two options are commonly used:

(a) Equivalent lateral force procedure and

(b) Dynamic analysis.

The selection of analysis method is based on the structure's seismic design category, structural system, dynamic properties and regularity, and economy. In most cases, the equivalent lateral force procedure is appropriate to determine lateral forces and their distribution. However, unusual structures with significant irregularities in shape, mass, or stiffness, or are affected by interaction with other structures, may require dynamic analysis. Table 12.6-1 of ASCE 7-16 shows the permitted analytical procedures for building structures in accordance with structural characteristics and seismic design categories. Section 15.1.3 of ASCE 7-16 indicates the permitted analytical procedures for nonbuilding structures. Appendix 4.G provides examples of configurations common to petrochemical facilities where dynamic analysis is recommended. Section 4.3.2 further describes various types of irregularities. If a dynamic analysis is performed, building codes also typically require earthquake base shear to be calculated by the equivalent static method and compared with that obtained from a dynamic analysis. Section 4.5.6 further discusses the background and the need to perform static analysis in addition to dynamic analysis when choosing the dynamic analysis option.

4.3.2 Equivalent Lateral Force Procedure

This analysis option allows the analyst to determine the seismic loads using static methods. Building codes typically contain restrictions on the use of the equivalent lateral force procedure for building-like structures, in particular when the structure has a highly irregular shape, nonuniform mass, or abrupt changes in lateral stiffness, or when it exceeds certain heights. Dynamic analysis is usually required to determine the distribution of lateral force in these situations. Ideally, mass distribution is uniform over the entire height of a structure, and the centroids of mass and rigidity coincide at every level. This ideal condition rarely occurs in petrochemical structures, but the distributions are in general close enough to justify the simplified code distribution of forces. The engineer should investigate the issue carefully before concluding that the equivalent static method is appropriate.

Tables 12.3-1 and 12.3-2 of ASCE 7-16 define horizontal and vertical structural irregularities typical of structures, respectively. Appendix 4.G provides examples of irregular petrochemical facilities where dynamic analysis is recommended in addition to the equivalent lateral force procedure. ASCE 7-16 Commentary Section C15.1.3 provides additional information on when dynamic analysis should be considered for nonbuilding structures.

4.3.3 Dynamic Analysis

This analysis procedure may be used for all petrochemical facility structures in any seismic design category with or without irregularities. However, this is the only permitted procedure for certain tall, flexible, and irregular structures assigned to Seismic Design Categories D, E, or F (see ASCE 7-16, Table 12.6-1 and Appendix 4.G).

4.4 EQUIVALENT LATERAL FORCE PROCEDURE

4.4.1 General

The concept of the equivalent lateral force procedure is to calculate the effective earthquake loads in terms of a base shear that is dependent on the structure's mass, the imposed ground acceleration, the structure's dynamic characteristics, the structure's inherent ductility, and the structure's importance. The base shear is then applied to the structure as an equivalent lateral load. This chapter also discusses how this lateral load should be distributed in plan and vertically along the height of the structure. Once this load is determined and distributed at various elevations of the structure, conventional static analysis techniques may be used to determine the seismic load in each structural member and at connections.

4.4.2 Determination of Base Shear and Seismic Load *E*

The total horizontal base shear (V) for a regular flexible building-like structure in any of the two orthogonal horizontal directions can be determined from Equation (4-1) (ASCE 7 Eq. 12.8-1):

$$V = C_S W \qquad (4-1)$$

where
   V = Base shear,
   W = Total seismic weight,
   C_S = Seismic response coefficient from Equation (4-2) (ASCE 7-16 Equation 12.8-2),

$$C_s = \frac{S_{DS}}{\left(\frac{R}{I_e}\right)} \qquad (4-2)$$

   S_{DS} = Design, 5% damped, spectral response acceleration parameter in the short-period range as determined from ASCE 7-16, Section 11.4.5,
   R = response Modification factors given in ASCE 7-16, Table 12.2-1 for buildings, ASCE 7-16, Table 15.4-1 for nonbuildings similar to buildings, and ASCE 7-16, Table 15.4-2 for nonbuildings not similar to buildings,
   I_e = Importance factor given in ASCE 7-16, Table 1.5-2,

Table 4-1. Importance Factor (I_e) for Different Occupancies.

Risk Category		Seismic Importance Factor for Structures, (I_e Factor)	Seismic Importance Factor for Components (I_p Factor)
(I or II) General	Normal	1.0	1.0
	Special	1^a	1^a
(III) Sufficient hazardous materials		1.25^b	1.50
(IV) Essential or containing a quantity of hazardous materials exceeding a prescribed threshold		1.5^b	1.50

[a]Quantification of importance factors for the "Special" usage category is an owner decision, because structures in this category are likely to be designed for higher seismic forces to protect an owner's investment.[b]Buildings containing hazardous materials and not classified as an essential facility may be classified as Category II if it can be demonstrated to the satisfaction of the Authority Having Jurisdiction that a release will not pose a threat to the public.

Section 15.4.1.1, and Table 4-1 of this book, and

S_S = Mapped MCE_R spectral response acceleration parameter at short periods as determined in accordance with Section 11.4.2.

The value of C_S need not exceed the following:

$$C_s = \frac{S_{DI}}{T\left(\frac{R}{I_e}\right)} \ for\ T \le T_L \tag{4-3}$$

$$C_s = \frac{S_{DI}T_L}{T^2\left(\frac{R}{I_e}\right)} \ for\ T > T_L \tag{4-4}$$

For building and nonbuilding structures similar to buildings, the minimum value of C_S shall not be less than the greater of $0.044S_{DS}I_e$ or 0.01. In addition, for structures located in areas where S_1 is equal to or greater than 0.6g, C_S should not be less than

$$C_s = \frac{0.5S_I}{\left(\frac{R}{I_e}\right)} \tag{4-5}$$

where

S_{D1} = Design, 5% damped, spectral response acceleration parameter at a period of 1.0 s, as determined from ASCE 7-16, Section 11.4.5,

T = Fundamental period of the structure (s) determined in ASCE 7-16, Section 15.4.4,

T_L = Long-period transition period (s) determined in ASCE 7-16, Section 11.4.6, and

S_1 = Mapped MCE$_R$ spectral response acceleration parameter at a period of 1 s as determined in accordance with ASCE 7-16, Section 11.4.2.

For nonbuilding structures not similar to buildings utilizing the R values from ASCE 7-16, Table 15.4-2, the minimum value of C_S shall not be less than the greater of $0.044S_{DS}I_e$ or 0.03. In addition, for those structures located in areas where S_1 is equal to or greater than 0.6g, C_S should not be less than

$$C_s = \frac{0.8S_1}{\left(\frac{R}{I_e}\right)} \tag{4-6}$$

Tanks and vessels are exempt from the minimums noted in the paragraph above if designed based on one of the following documents: AWWA D100 (2011), AWWA D103 (2009), API 650 Appendix E (2014), and API 620 Appendix L (2014, as modified by ASCE 7). If the tanks and vessels are designed to one of these documents, the minimum value of C_s is the same as that required for nonbuilding structures similar to buildings noted above.

Each parameter in Equations (4-1) through (4-6) represents a certain aspect of the earthquake loading, described as follows:

S_1 and S_S are the mapped MCE, 5% damped, site class B, spectral response acceleration parameters at a period of 1 s and 0.2 s, respectively. They represent an estimate of the maximum seismic motion expected at the site and are therefore site dependent. For cases where site-specific data exist, using the site-specific spectrum in lieu of those given by the code of record, subject to the approval of the local building official, is appropriate.

Values for S_1 and S_S are based on risk-adjusted (2% in 50 years) spectral response accelerations targeted to achieve building designs with a 1% probability of collapse in 50 years (i.e., uniform risk). Site-specific values for S_1 and S_S should be based on the site's coordinates (longitude, latitude). The ASCE 7 Hazard Tool (found at https://asce7hazardtool.online/) can be used to look up ground motion values, in addition to other hazard data. Values based on site coordinates should be used for design because the maps provided with ASCE 7 cannot be read accurately in high seismic areas. Values based on zip codes should not be used because the values can vary significantly within a given zip code. In general, S_1 and S_S are modified by the site coefficients, F_V and F_A, respectively to arrive at S_{M1} and S_{MS}. F_V, the velocity-related soil factor, and F_A, the acceleration-related soil factor are used to account for local soil conditions. The soil conditions are separated into six site classes: A, B, C, D, E, and F. Unfortunately, the shape of the ASCE 7 response spectrum on soft sites has an unconservative problem governed by large-magnitude events. For structures on Site Class E sites with S_s

greater than or equal to 1.0 and structures on Site Class D and E sites with S_1 greater than or equal to 0.2, ASCE 7-16, Section 11.4.8 requires that a site-specific ground motion hazard analysis be performed. While ASCE 7-16, Section 11.4.8 contains exceptions to these requirements, the user will find that the exception may be quite conservative and that a site-specific ground motion hazard analysis may prove to be more economical. Also note that to apply the exceptions to ASCE 7-16, Section 11.4.8, the user must apply Supplement 1 to ASCE 7-16. The ASCE 7 seismic design provisions are expected to provide a low likelihood of collapse against an earthquake with at least 1.5 times larger ground motions than the maximum site-adjusted design ground motion. Consequently, S_{M1} and S_{MS} are multiplied by 2/3 (1/1.5) to arrive at the final design values S_{D1} and S_{DS}. The foregoing is expressed in the following Equations (4-7) through (4-10):

S_{MS} = MCE$_R$, 5% damped, spectral response acceleration parameter at

  short periods adjusted for site class effects as defined in Section 11.4.4

$$= F_a S_S \qquad (4\text{-}7)$$

S_{M1} = MCE$_R$, 5% damped, spectral response acceleration parameter at a

  period of 1 s adjusted for site class effects as defined in Section 11.4.4

$$= F_v S_1 \qquad (4\text{-}8)$$

$$S_{DS} = \frac{2}{3} S_{MS} \qquad (4\text{-}9)$$

$$S_{D1} = \frac{2}{3} S_{M1} \qquad (4\text{-}10)$$

C_S, in Equations (4-1) through (4-6), represents the spectral acceleration corresponding to the period of the structure and accounts for potential amplification of seismic forces due to site-specific soil conditions, ductility, and overstrength of the structure. C_S depends on six terms: S_{DS}, S_{D1}, T, T_L, R, and I_e.

I_e is a measure of the structure's relevant importance. For more important structures, higher seismic forces are prescribed. The seismic importance factor, I_e, provides a means of increasing the design force levels for facilities that may have an unusual hazard, have a potential for releasing hazardous materials, or are important for emergency response. The I_e value for individual petrochemical facilities should be reviewed with the owner and/or the building official for those cases where items do not fall under the code definitions. Table 4-1 gives suggested values of I_e for petrochemical facilities. Note that an entire facility need not use the same value of I_e. See Chapter 2 for further discussion on this topic.

Total seismic weight, W, includes the total dead load of the structure; weights of attached fixed process equipment and machinery, piping, valves, electrical cable trays, and the contents of these items; and code-specified portions of live and snow load. Generally, the only live load that need be considered is a minimum of 25% of the floor live load in areas used for storage. Live load, except as mentioned above, is not included in the seismic weight because it is not attached to the structure and therefore does not contribute to the structure's inertia. Also, in nonbuilding structures, most design live loads represent maintenance loads that occur for a short duration. Therefore, the probability of maintenance loads being present during the design seismic event is very low. The amount of snow load to include in the seismic weight differs depending on whether the structure is a building or a nonbuilding structure. For a building, 20% of the uniform design snow load, regardless of actual roof slope, is used where the flat roof snow load, P_f, exceeds 30 lb/ft². For a nonbuilding structure, the seismic weight includes snow and ice loads where these loads constitute 25% or more of W or where required by the Authority Having Jurisdiction based on local environmental characteristics.

The response modification factor, R, incorporates the combined effects of the inherent ductility and the overstrength capacity of the structure. R reduces the seismic design forces to a level meant to be used in an elastic analysis, but still achieves an acceptable inelastic performance. Ductility is the structure's ability to dissipate energy as it vibrates back and forth, particularly in the inelastic range (i.e., hysteretic behavior) and depends on both the type of structure and the structure's design details. Overstrength comes from several sources: actual material strengths larger than prescribed minimums, successive yielding for noncritical portions of the structure, inherent overdesign, the ϕ factor incorporated in the design practices, and prescribed story drift limits. For a more detailed explanation of this term, refer to the ASCE 7 commentary.

The R factors for buildings, nonbuilding structures similar to buildings, and nonbuilding structures not similar to buildings can be found in ASCE 7 Tables 12.2-1, 15.4-1, and 15.4-2, respectively. R values for nonbuilding structures in ASCE 7 Table 15.4-2 are generally smaller than those for buildings. This is because buildings tend to have structural redundancy due to multiple bays and frame lines. Also, buildings contain nonstructural elements, which effectively give the building greater damping and strength during strong ground motion response. Note also that using higher R values in design tends to result in a flexible structure that tends to increase calculated displacements.

ASCE 7 Equation (12.8-7) is for buildings with uniform mass and stiffness and should not be used for nonbuilding structures. For nonbuilding structures, the following formula can be used to calculate the fundamental period. The formula is based on Rayleigh's method utilizing the structural properties and deformational characteristics of the structure as determined by a static analysis. Many commonly used structural analysis programs have the Rayleigh method procedure built in as a way of calculating fundamental periods for nonbuilding structures.

$$T = 2\pi \sqrt{\frac{\sum\limits_{i=1}^{n} w_i \delta_i^2}{g \sum\limits_{i=1}^{n} f_i \delta_i^2}} \qquad (4\text{-}11)$$

where

f_i = Seismic lateral force at level i,
w_i = Sravity load at level i,
δ = Elastic static displacement at level i due to the forces f_i, and
g = Acceleration of gravity.

Alternatively, the fundamental period of the structure may be estimated by a frequency (modal) analysis. The engineer should use judgment as to which method is most desirable for the application at hand. Appendix 4.A presents example calculations of the fundamental period of vibration, T, for several typical structures commonly found in petrochemical facilities.

For buildings, ASCE 7 places an upper limit on the period determined by analysis. The fundamental period, T, is limited to the product of the coefficient for upper limit on calculated period (C_u) from ASCE 7-16, Table 12.8-1 and the approximate fundamental period Ta determined using ASCE 7-16, Equation (12.8-7). This limit is imposed on the period determined by analysis, because it tends to be less than the period of the actual structure and can lead to an unconservative design. ASCE 7 does not impose these same limits for nonbuilding structures similar to buildings or nonbuilding structures not similar to buildings.

Figure 4-1 illustrates the limits placed on the calculation of the seismic response coefficient, C_S, given by Equations (4-1) through (4-6).

where

T = Fundamental period of the structure, s,
$T_S = S_{D1} / S_{DS}$,
$T_0 = 0.2T_S$, and
T_L = Long-period transition period, seconds.

Appendixes 4.B and 4.C provide examples of determination of base shear for typical petrochemical structures. For nonbuilding structures that are rigid ($T < 0.06$ s), such as a short, stiff support piers for a horizontal vessel, the base shear may be calculated by Equation (4-12) [ASCE 7-16, Equation (15.4-5)].

$$V = 0.30 S_{DS} W I_e \qquad (4\text{-}12)$$

The seismic load on the structure, E, is made up of two components: E_h and E_v. E_h is the horizontal seismic effect calculated from the base shear (V or F_p) and modified by the redundancy factor ρ. E_v is the vertical seismic effect calculated from a percentage of the design spectral response acceleration parameter, S_{DS},

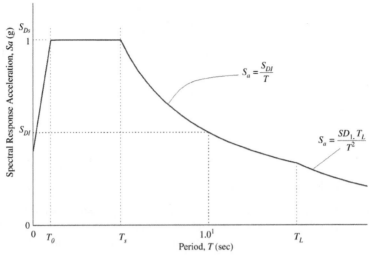

Figure 4-1. Typical design response spectrum.

times the dead load of the structure. The foregoing is expressed in the following Equations (4-13), (4-14), (4-15a), and (4-15b):

$$E = E_h + E_v \qquad (4\text{-}13)$$

$$E_h = \rho Q_E \qquad (4\text{-}14)$$

where ρ is meant to reflect the amount of redundancy in the SFRS in each direction applicable to the structure and Q_E is the effect of horizontal seismic forces.

$$E_v = 0.2 S_{DS} D \qquad (4\text{-}15a)$$

where D is the dead load of the structure.

Exception: Where the option to incorporate the effects of vertical seismic ground motions using the provisions of Section 11.9 is required in ASCE 7-16, Section 15.1.4, the vertical seismic load effect, E_v, shall be determined in accordance with Equation (4-15b) as follows:

$$Ev = 0.3 S_{av} D \qquad (4\text{-}15b)$$

where S_{av} is the design vertical response spectral acceleration determined in ASCE 7-16 Section 11.9.

ASCE 7-16, Section 15.1.4 requires that tanks, vessels, hanging structures, and nonbuilding structures incorporating horizontal cantilevers use the vertical response spectrum defined in ASCE 7-16, Section 11.9.

The value of ρ is assigned as either 1.0 or 1.3 using the rules of ASCE 7-16, Section 12.3.4 and Section 12.3.4.1 allows ρ to equal 1.0 for

- Structures assigned to Seismic Design Categories B and C;
- Design of structures not similar to buildings;
- Design of collector elements, splices, and their connections;
- Drift calculations and P-delta effects;
- Design of nonstructural elements;
- Diaphragm loads;
- Members or connections where the system overstrength effect, Ωo has been considered;
- Structures with damping systems designed to ASCE 7 Chapter 18; and
- Design of structural walls for out-of-plane forces, including their anchorage.

Properly designed and detailed structures can resist seismic loads two to four times their prescribed forces due to structural redundancy, progressive yielding, and actual material strengths. To ensure the inelastic response of the structure during a design event, critical portions of the structure and its connections are required to be designed for the system overstrength effect, E_{mh} ASCE 7-16, Equations (12.4-5) and (12.4-6) define the seismic load effect with overstrength as

$$E = E_{mh} +/- E_v \qquad (4\text{-}16)$$

where

E_{mh} is the horizontal seismic forces including the structural overstrength and

$$E_{mh} = \Omega o Q_E \qquad (4\text{-}17)$$

where

Ωo is the overstrength factor found in ASCE 7-16, Tables 12.2-1, 15.4-1, and 15.4-2 and ranges from 1.0 to 3.0. The overstrength factor is dependent on the structural resisting system.

4.4.3 Combination Structures

Combinations of structural systems are often incorporated into the same structure. Structures that include more than one structural system require special attention. Section 15.3 of ASCE 7-16 provides the procedures for determining seismic loads on combination structures. The following summarizes those procedures.

Nonbuilding structures supported by other structures, where the weight of the nonbuilding structure is *less than 25%* of the combined weight of the nonbuilding structure and the supporting structure, are to be designed for seismic forces determined according to the provisions of Chapter 13 of ASCE 7. The use of Chapter 13 treats the nonbuilding structure as a nonstructural component. The supporting structure is to be designed for seismic forces using the provisions of ASCE 7-16, Section 12 or Section 15.5, as appropriate, with the weight of the nonbuilding structure considered to be part of the seismic weight, W. In other words, the nonbuilding structure is treated as just another mass in the supporting structure for the design of the supporting structure. Although ASCE 7-16 allows the a_p and R_p values from Chapter 13 to be used for design, in some cases these values are unconservative. For example, where an item such as a vessel is supported on a structural frame that is in turn supported by a large nonbuilding structure, the value of a_p should be taken as 2.5 and not as 1.0 because few, if any, of the nonstructural components listed in Chapter 13 meet the definition of rigid.

Nonbuilding structures supported by other structures, where the weight of the nonbuilding structure is *greater than or equal to 25%* of the combined weight of the nonbuilding structure and the supporting structure, are to be designed for seismic forces determined using an analysis that combines the structural characteristics of both the nonbuilding structure and the supporting structure. The design rules, described as follows, differ if the nonbuilding structure can be considered a rigid element or a flexible (nonrigid) element.

1. Where the nonbuilding structure can be considered a *rigid component* (period less than 0.06 s), the nonbuilding structure and its attachments are to be designed as nonstructural components (ASCE 7-16, Chapter 13) for seismic forces determined using a value of R_p equal to the R value of the nonbuilding structure from ASCE 7-16, Table 15.4-2 and an a_p value equal to 1.0. The supporting structure is to be designed for seismic forces as it normally would be designed using the provisions of ASCE 7-16, Section 12 or 15.5, as appropriate, with the weight of the nonbuilding structure considered to be part of the seismic weight, W.

2. Where the nonbuilding structure can be considered a *flexible component* (period greater than or equal to 0.06 s), the nonbuilding structure and supporting structure are to be modeled together in a combined model with the appropriate stiffness and seismic weight distribution. The combined structure is to be designed in accordance with ASCE 7-16, Section 15.5, with the R value of the combined system taken as the lesser of either the R value of the nonbuilding structure or the R value of the supporting structure. The nonbuilding structure and attachments are to be designed for the forces determined for the nonbuilding structure in the combined analysis.

Another class of combination structure not specifically addressed in ASCE 7 is one in which the nonbuilding structure is capable of carrying vertical load but relies on an independent structure to partially or completely resist overturning from horizontal wind and seismic loads. An example of this type of combination structure is a vertical stack enclosed by a vertical space frame. The supporting space frame and vertical stack are to be modeled together in a combined model with the appropriate stiffness and seismic weight distribution. The combined structure is to be designed in accordance with ASCE 7-16, Section 15.5, with the R value of the combined system taken as the lesser of either the R value of the nonbuilding structure or the R value of the supporting structure. The nonbuilding structure and attachments are to be designed for the forces determined for the nonbuilding structure in the combined analysis.

Appendix 4.B contains guidelines for determining base shear for nonbuilding structures that are a combination of more than one structural system (e.g., finfans supported on pipeways). Specific guidance is provided for combinations of rigid and nonrigid structures and for supported equipment weighing greater than and less than 25% of the combined weight of the nonbuilding structure and the supporting structure. Appendix 4.C also contains examples showing calculations of base shear for combination structures (e.g., examples 1.2 and 3.2b).

4.4.4 Vertical Distribution of Forces

Lateral force distribution in the vertical direction is a method for resolving the base shear into static force equivalents applied laterally to the structure. When the mass of the structure is uniformly distributed over the height of the structure (as is the case in buildings), this force distribution is assumed to be linear (inverted triangular) for structures with a period less than or equal to 0.5 s, parabolic for structures with a period greater than or equal to 2.5 s, and a transitional shape between linear and parabolic for structures with periods between 0.5 and 2.5 s. For multilevel buildings and structures, the base shear, V, is distributed to different levels assuming a distribution of forces as described above. This is a close representation of the first-mode shape of the structures. It also is representative of the average of the shear deflection and moment deflection curve for the fundamental modes. ASCE 7-16, Section 12.8.3 presents an example of a methodology to distribute lateral forces in such a way. Figure 4-2 shows the possible distributions.

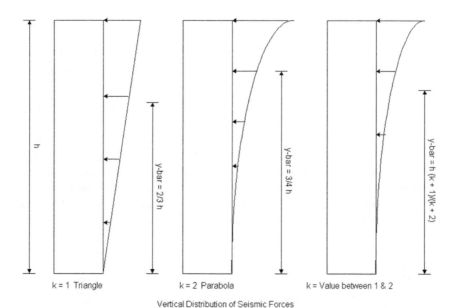

Vertical Distribution of Seismic Forces

Figure 4-2. Vertical distribution of seismic forces.

When the structure under consideration is highly irregular in shape and/or has nonuniform mass distribution vertically, the concept of lateral load distribution as previously defined is no longer valid. In such cases, a proper dynamic analysis should be conducted to determine lateral load distribution.

4.4.5 Horizontal Distribution of Forces

Once horizontal forces for each elevation are determined, they are distributed to each of the horizontal load-carrying elements in the ratio of their horizontal stiffnesses. Such an approach assumes that diaphragms are rigid. For structures with flexible diaphragms, the horizontal stiffness of the diaphragm should be incorporated when distributing horizontal forces into various load-carrying elements. Alternatively, the lateral forces should be assigned based on tributary spans. ASCE 7-16, Section 12.8.4 presents an example methodology for horizontal distribution of forces.

4.4.6 Torsional Effects

Horizontal torsional moments occur in structures with rigid diaphragms as a result of eccentricities between the center of mass and the center of rigidity at each elevation. Other factors also contribute to horizontal torsion, including

(a) SFRSs that are located primarily near the center of the structure rather than the perimeter,

(b) Disproportionate concentration of inelastic demands in system components,

(c) The effects of nonstructural elements,

(d) Uncertainties in defining the structure's stiffness characteristics, and

(e) Spatial variation (and rotational components of ground motions) of horizontal input motions applied to long structures.

Where torsional irregularities exist, the effect of horizontal torsion can be computed at each floor level and included in the analysis in accordance with requirements of building codes.

ASCE 7-16, Section 15.4.1(5) provides relief from designing for accidental torsion for most nonbuilding structures. Provided that the mass locations for the structure, any contents, and any supported structural or nonstructural elements (including, but not limited to, piping and stairs) that could contribute to the mass or stiffness of the structure are accounted for and quantified in the analysis, the following accidental torsion requirements of ASCE 7-16, Section 12.8.4.2 need not be accounted for:

(a) Rigid nonbuilding structures, or

(b) Nonbuilding structures not similar to buildings designed with R values less than or equal to 3.5, or

(c) Nonbuilding structures similar to buildings with R values less than or equal to 3.5, provided that one of the following conditions is met:

    i. The calculated center of rigidity at each diaphragm is greater than 5% of the plan dimension of the diaphragm in each direction from the calculated center of mass of the diaphragm, or

    ii. The structure does not have a horizontal torsional irregularity type 1A or 1B and the structure has at least two lines of lateral resistance in each of two major axis directions. At least one line of lateral resistance shall be provided at a distance of not less than 20% of the structure's plan dimension from the center of mass on each side of the center of mass.

In addition, structures designed to this section should be analyzed using a three-dimensional representation in accordance with Section 12.7.3.

Note that the mass amount and location can change during operations, and the analysis must account for this change. An example of a change would be the emptying and loading of batch coker vessels, which will change both the location and amount of mass in the structure.

Accidental torsion is typically accounted for by assuming the center of mass is shifted in each horizontal direction from its calculated value by a distance equal to 5% of the structure dimension perpendicular to the direction being considered. Sections 12.8.4.2 and 12.8.4.3 of ASCE 7-16 provide an example of such a methodology.

4.4.7 Overturning Potential

Every structure should be designed and evaluated against potential overturning effects caused by earthquake forces. The overturning moment to be resisted at each level should be determined using the seismic forces that act on levels above the level under consideration. Overturning effects on every element should be carried down to the foundations.

Chapter 5 discusses safety factors to be used to check for potential against foundation overturning for new facilities. In addition, Section 4.6 presents special provisions that may be used to evaluate for foundation stability against overturning for existing facilities.

4.4.8 Sliding Potential

Building codes typically do not require any checks for stability against sliding, because building-like structures are required to be properly secured at the foundation. Likewise for the design of new petrochemical facilities, almost all structures (with the possible exception of large-diameter flat bottom storage tanks) should be anchored to their foundations.

Chapter 5 discusses safety factors to be used to check for potential against sliding, if needed, for new facilities. Section 4.6 outlines requirements for checking stability against sliding for structures in existing facilities that are not anchored to their foundations.

4.4.9 Directions of Earthquake Forces

The directions of earthquake forces used in the design of structures are those that will produce the most critical load effects in the structures. Depending on the structure's seismic design category, this requirement may be satisfied as follows.

For structures in Seismic Design Category B, ASCE 7 allows the design seismic forces to be applied independently in each of the two orthogonal directions without consideration of orthogonal interaction effects.

The same holds true for structures in Seismic Design Category C unless the structure has a horizontal structural irregularity Type 5 (nonparallel system). If a horizontal structural irregularity Type 5 exists, the requirement that orthogonal effects be considered may be satisfied by designing such elements using the following method:

> Consider 100% of the member forces due to loads applied in one horizontal direction and 30% of the member forces due to loads applied in the orthogonal horizontal direction. The intent of this procedure is to approximate the most critical load effect in the components of the structure without having to evaluate the structure for loads in any direction.

For structures in Seismic Design Categories D through F, the requirements of Seismic Design Category C must be met. Additionally, any column or wall that forms part of two or more intersecting seismic force–resisting systems (e.g., corner columns) and is subject to axial load exceeding 20% of the axial design strength of the column or wall must consider the orthogonal interaction effects. This requirement may be satisfied by designing such elements using the following method:

> Consider 100% of the member forces due to loads applied in one horizontal direction and 30% of the member forces due to loads applied in the orthogonal horizontal direction. The intent of this procedure is to approximate the most critical load effect in the components of the structure without having to evaluate the structure for loads in any direction.

As mentioned, use of the 100-30 rule is intended to approximate the most critical load effects in the elements making up the structure. For symmetrical nonbuilding structures such as column-supported tanks, vertical vessels, and ground storage tanks, the intent of ASCE 7 is met by applying the earthquake force in one direction for a vertical vessel or ground storage tank or by applying the earthquake force at the column and halfway between the columns of a column-supported tank.

4.4.10 Vertical Motions

Response effects due to vertical components of ground motion are not required to be calculated where the nonbuilding structure is assigned to Seismic Design Category B or where determining demands on the soil-structure interface of foundations. They are considered to be allowed for by the following special provision:

> Combined vertical gravity loads and horizontal seismic forces must be considered in designing building components. Uplift effects due to seismic forces must also be considered. For materials designed using working stress procedures, dead loads are multiplied by 0.9 when used to check against uplift in the IBC Alternate Basic Load Combinations.

Note that for facilities that are located in areas of known high seismicity and close to active faults, close attention must be given to near-field effects. Near-field terminology is used when the epicenter of the earthquake is close to a site of interest (see Chapter 3). Because compression waves attenuate much faster than horizontal shear waves, the vertical component of the earthquake tends to be dominant only in the near field. For these special cases, the designer/analyst should give appropriate consideration to the effects of vertical accelerations on design.

Where the nonbuilding structure is assigned to Seismic Design Categories C through F, vertical earthquake forces are to be considered explicitly in the design or analysis of the structure. The vertical seismic load effect is to be taken as $0.2S_{DS}D$, where D is the effective seismic weight, or as $0.3S_{av}D$ for tanks, vessels, hanging structures, and nonbuilding structures incorporating horizontal cantilevers. Alternate factors may be used if substantiated by site-specific data. Once the vertical ground motion is defined, appropriate consideration should be given to the structure's dynamic response in the vertical direction. Member forces and moments due to the vertical component of the earthquake should be obtained and combined with the forces and moments resulting from the horizontal component of the earthquake in accordance with the procedure of Section 4.4.9. Refer to Chapter 3 for more guidance on the definition of vertical earthquake motion.

In addition, for Seismic Design Categories D through F, horizontal cantilever structural components are to be designed for a minimum net upward force of 0.2 times the dead load for buildings and $0.3S_{av}D$ for nonbuilding structures in addition to the applicable load combinations.

4.4.11 Nonstructural Components

Nonstructural components including architectural components are defined as elements, equipment, or components permanently attached to the structure under consideration. As a general guideline, nonstructural components are those with weights less than 25% of the supporting structure's effective seismic weight as defined in Section 12.7.2 of ASCE 7-16. Any component or structure founded directly on soil/ground or sharing the foundation of another structure is not considered a nonstructural component. Also, the following nonstructural components are exempt from the requirements of this section:

1. Mechanical and electrical components in Seismic Design Category B.

2. Mechanical and electrical components in Seismic Design Category C, provided that either of the following conditions apply:

(a) Component importance factor (I_p) is equal to 1.0; or

(b) Components weigh 20 lb (89 N) or less, or the weight of a distributed system is 5 lb/ft (73 N/m) or less.

3. Discrete mechanical and electrical components in Seismic Design Categories D, E, or F that are positively attached to the structure, provided that either the

(a) Component weighs 400 lb (1,779 N) or less; the center of mass is located 4 ft (1.22 m) or less above the adjacent floor level; flexible connections are provided between the component and associated ductwork, piping, and conduit; and the component importance factor, I_p, is equal to 1.0; or

(b) Component weighs 20 lb (89 N) or less, or the weight of a distributed system is 5 lb/ft (73 N/m) or less.

4. Distribution systems in Seismic Design Categories D, E, or F included in the exceptions for conduit, cable tray, and raceways in ASCE 7-16, Section 13.6.5, duct systems in ASCE 7-16, Section 13.6.6, and piping and tubing systems in ASCE 7-16, Section 13.6.7.3. Where in-line components, such as valves, in-line suspended pumps, and mixing boxes, require independent support, they shall be addressed as discrete components and shall be braced considering the tributary contribution of the attached distribution system.

Nonstructural components should be designed in accordance with ASCE 7, Chapter 13, with the following considerations:

1. Nonstructural components should be assigned to the same seismic design category as the structure that they occupy or to which they are attached.

2. Components should be assigned an importance factor, I_p, equal to 1.0 unless any of the following conditions exist, then the importance factor, I_p, should be taken as 1.5.

   • Component is required to function for life-safety purposes after an earthquake, including fire protection sprinkler systems and egress stairways.

   • Component conveys, supports, or otherwise contains toxic, highly toxic, or explosive substances where the quantity of the material exceeds a threshold quantity established by the Authority Having Jurisdiction and is sufficient to pose a threat to the public if released. The component is in or attached to a Risk Category IV structure and is needed for continued operation of the facility, or its failure could impair the continued operation of the facility.

   • Component conveys, supports, or otherwise contains hazardous substances and is attached to a structure or portion thereof classified by the Authority Having Jurisdiction as a hazardous occupancy.

3. Testing or the use of experience data should be deemed an acceptable alternative method to the analytical method to determine the seismic capacity of the components and their supports and attachments for components assigned an $I_p = 1.0$.

4. For components assigned an $I_p = 1.5$, the requirements of ASCE 7-16, Section 13.2.2 are to be followed, including the requirement that active mechanical and electrical equipment must be seismically qualified by testing or experience data when the structures/facilities they are associated with have an assigned Seismic Design Category of C or higher.

5. FEMA 74 (2012), *Reducing the Risks of Nonstructural Earthquake Damage*, may be used for less significant nonstructural components such as heavy bookshelves, individual computers, and so on, to prevent their interaction with or impact upon nearby critical nonstructural components (i.e., becoming flying objects during an earthquake).

4.5 DYNAMIC ANALYSIS METHODS

4.5.1 General

Dynamic analyses of structures are typically conducted using structural computer programs. Dynamic analysis can be performed for any type of structure at the engineer's discretion. However, it must be performed in cases where certain structural irregularities exist.

Current availability of high-performance computers at low prices and availability of various structural analysis software packages have made the option of dynamic analysis increasingly attractive to engineers. In general, dynamic analyses are believed to result in more accurate determination of a structure's dynamic properties and hence its response to a seismic event. Dynamic analyses better capture local effects in addition to global response, often leading to a more economical design of a new facility and avoiding unnecessary conservatism in evaluating an existing facility.

In choosing the dynamic analysis option, the engineer should have a good understanding of the structure's global behavior and should review the dynamic analysis results carefully, to ensure response is as expected. This is often achieved by comparing the base shear with that obtained using the equivalent static approach.

4.5.2 Ground Motion

The ground motion to be used in a dynamic analysis is defined in terms of either a ground response spectrum (most common) or a time history. In general, the ground response spectrum is first defined. Chapter 3 describes the logic of the choice of ground motion and its development. The response spectrum shape is either a standard shape scaled to an appropriate peak ground acceleration or is a site-specific shape. The spectral shape for new design typically corresponds to two-thirds of a risk-targeted MCE_R response spectrum based on a 10% probability of collapse exceedance for MCE-level ground motions and is generally defined for several damping ratios. For petrochemical-type structures, analysis is usually performed using the 5% damped spectral shape; therefore, as a minimum, the ground response spectra should be defined for 5% damping.

If a time history (referred to as "response history" in ASCE 7) analysis methodology is chosen, the ground motion must be defined in terms of time history of response (usually acceleration, sometimes velocity or displacement). Either actual earthquake time histories or time histories generated artificially to represent a spectral shape are suitable for use. However, in both cases, the response spectrum associated with the chosen time histories should match the site-specific response spectrum at the frequencies of interest. Chapter 3 discusses the development of the ground (free-field) time history for use in dynamic analysis.

4.5.3 Mathematical Model

A finite element or lumped mass model of the structure is a mathematical representation of the structure that properly allows for the distribution of the structure's mass and stiffness to an extent that is adequate for calculating the significant features of its dynamic response.

To simplify the analysis, neglecting the soil-structure interaction effect (see Section 4.5.7) and assuming the structures to be fixed at the base is permitted by code.

A three-dimensional structural model should be used for the dynamic analysis of structures with highly irregular plan configurations, such as those having a plan irregularity defined in Table 12.3-1 of ASCE 7-16 and having a rigid diaphragm. For regular structures, use of a two-dimensional model is usually adequate. In general, two two-dimensional models, each representing an orthogonal principal axis of the structure, are developed and used in the analysis.

Structures with a dynamic response that can be characterized as either a shear-type response or a bending-type response can be reasonably modeled using beam elements. Three-dimensional finite element models utilizing plate/shell or three-dimensional solid elements may be used for more complex structures.

4.5.4 Response Spectrum Analysis

A response spectrum analysis is an elastic dynamic analysis of a structure that utilizes the peak dynamic response of significant modes that contribute to total structural response. Peak modal responses are calculated using the ordinates of the input spectrum, which correspond to modal periods. Maximum modal responses are then combined to obtain the peak structural response.

The following considerations should be given when performing a response spectrum analysis:

(a) Number of modes: The analysis should include enough modes to obtain a combined modal mass participation of 100% of the structure's mass. For this purpose, representing all modes with periods less than 0.05 s in a single rigid body mode with a period of 0.05 s is permitted. Alternatively, the analysis may include a minimum number of modes to obtain a combined mass modal participation of at least 90% of the actual mass in each orthogonal direction of response considered in the model.

(b) Combination of modes: In a response spectrum analysis option, the maximum response of the structure during each mode is first computed. These modal responses must be combined to obtain the peak response of the structure. Modal responses are generally combined by the square root sum of the squares (SRSS) method. However, the SRSS method may be unconservative for structures with closely spaced modes. For this case the complete quadratic combination(CQC) method should be used.

(c) Design values: Where the design value for the modal base shear (Vt) is less than 100% of the calculated base shear (V) using the equivalent lateral force procedure, the design forces shall be multiplied by the following modification factor (see Section 4.5.6): V/Vt. For distributed mass cantilever structures listed in ASCE 7-16, Table 15.4-2, including steel and reinforced concrete stacks, chimneys, silos, skirt-supported vertical vessels, and steel tubular support structures for onshore wind turbine generator systems using the modal analysis procedure of ASCE 7-16, Section 12.9.1 and the combined response for the modal base shear (Vt) at less than 85% of the calculated base shear (V) using the equivalent lateral force procedure, multiplying the forces by $0.85V/Vt$ is permitted in lieu of the provisions of ASCE 7-16, Section 12.9.1.

4.5.5 Time History Analysis

A time history analysis (also known as a response history analysis) is generally a more complex and expensive method of performing a dynamic analysis, but is considered to give a more realistic estimate of structural response to a given earthquake than a response spectrum analysis can. In a time history approach, the input is defined in terms of time histories of ground motion. Modal and directional combinations are performed in the time or frequency domains, accounting for duration, frequency content, and phasing for a particular earthquake time history.

In general, two options are available for performing time history analyses: the modal superposition and the direct integration schemes. A modal superposition approach is similar in concept to the response spectrum approach, with the exception that individual modal responses are obtained as a time history and, as such, can be combined with other modal responses in an algebraic sense in the time domain. Hence, the phasing between modes is retained. The direct integration scheme determines the total response of the structure at each time step by solving the equations of motion using direct integration.

The discussion above concerns linear response history analysis. In rare occasions, nonlinear dynamic analysis may need to be performed. Although this option is rarely used in practice for petrochemical facilities, special circumstances might warrant a nonlinear analysis. Only the direct integration scheme can be used in nonlinear dynamic analysis. In performing nonlinear analysis, special consideration should be given to actual structural nonlinearity and the structure's nonlinear hysteretic behavior in contrast to the ductility values (R factors) that are used as means of energy dissipation capability of a structure or structural system.

Similar to the response spectrum analysis, in a time history modal superposition analysis the analysis should include enough modes to obtain a combined modal mass participation of 100% of the structure's mass. For this purpose, representing all modes with periods less than 0.05 s in a single rigid body mode with a period of 0.05 s is permitted. In a direct integration scheme, by definition, 100% of the mass is considered in the solution at each time step.

4.5.6 Scaling of Results

When a dynamic analysis is performed, various building codes generally require that the base shear (after adjusting by I_e/R) be compared with that of an equivalent static analysis. According to ASCE 7-16, Section 12.9.1.4, if the calculated dynamic base shear from a response spectrum analysis is less than 100% of the equivalent static base shear, it should be scaled up such that the dynamic base shear equals 100% of that obtained using the equivalent lateral force procedure. ASCE 7-16, Section 15.1.3 allows a reduced scaling for distributed mass cantilever structures such as stacks, chimneys, silos, skirt-supported vertical vessels, and steel tubular support structures for onshore wind turbine generator systems. Using the modal analysis procedure of Section 12.9.1 and the combined response for the modal base shear (V_t) at less than 85% of the calculated base shear (V) using the equivalent lateral force procedure, multiplying the forces by $0.85V/V_t$ is permitted in lieu of the provisions of Section 12.9.

Long recognized is that the equivalent lateral force method using an R factor was never intended to provide an accurate prediction of the structural loads when the structure becomes inelastic. Rather, it was intended to provide an acceptable margin against collapse when using a linear analysis method. While using the response spectrum method may more accurately predict structural response and force distribution, it is still a linear elastic analysis that is not intended to accurately predict inelastic behavior and structural response. The resulting forces and moments should not be considered to be any more accurate than those resulting from an equivalent lateral force analysis when the structure behaves in a nonlinear manner. Thus, ASCE 7-16 has been changed to require that results from a modal analysis be scaled so that the base shear equals 100% of the results from the equivalent lateral force analysis to provide an equivalent margin of safety against collapse when using linear rather than nonlinear methods.

4.5.7 Soil-Structure Interaction

Soil-structure interaction (SSI) refers to dynamic interaction effects between a structure and the underlying soil during a seismic event. SSI effects are pronounced for heavy embedded structures founded on soft or medium soil and in general are negligible for light and surface-founded structures, or structures founded on competent material (stiff soil or rock).

For structures founded on rock or very stiff soils, the foundation motion is essentially that which would exist in the soil at the level of the foundation in the absence of the structure and any excavation; this motion is denoted the free-field ground motion. For soft soils, the foundation motion differs from that in the free field due to the coupling of the soil and structure during the earthquake. This interaction results from the scattering of waves from the foundation and the radiation of energy from the structure due to structural vibrations. Because of these effects, the state of deformation (particle displacements, velocities, and accelerations) in the supporting soil differs from that in the free field. In turn, the dynamic response of a structure supported on soft soil may differ substantially in amplitude and frequency content from the response of an identical structure supported on a very stiff soil or rock.

SSI effects usually result in potential de-amplification of the structural response, depending on the site-specific conditions and the combined soil and structure dynamic characteristics. Generally, a shift tends to occur in the combined soil-structure natural period as compared with the period of the structure assumed to be fixed at the base. This shift tends to increase the system period.

Methodologies for considering SSI effects are presented in ASCE 7 (Chapter 19) and are available in the literature (Wolf 1985). These effects can simply be allowed for, if the underlying soil happens to be a uniform medium and as such can be represented by an equivalent spring and a damper. If the underlying soil is layered, properly allowing for the effects of SSI requires more complicated techniques such as finite element analysis, or other alternate methods.

As a rule of thumb, when SSI effects are profound, the following tends to happen:

(a) The system period increases.

(b) An overall reduction occurs in structural response (unless the structure and the underlying soil happen to be in resonance, e.g., Mexico City, mid-rise buildings).

(c) More flexible subsystems may have increased response and should be given more attention in the seismic design.

(d) The rocking effects are more profound resulting in higher structural displacements.

4.6 CONSIDERATIONS FOR EXISTING FACILITIES

4.6.1 General

The procedures outlined above apply primarily to the design of new facilities. This section provides specific guidance for seismically evaluating existing petrochemical facilities. Different philosophies and methodologies are appropriate for evaluating existing facilities because

(a) Older facilities were designed to earlier codes and standards, wherein the effect of earthquakes on structures and the structure behavior were not adequately studied and understood.

(b) While the objective is to improve seismic safety of existing facilities, upgrading existing facilities to be in compliance with the requirements of current codes and standards would be very difficult.

(c) Designers can be more conservative in new designs. However, having the same consideration for retrofitting existing structures is rather difficult because not only do designers have to meet a specific set of seismic upgrade criteria but they also deal with other issues such as limited spacing, interference, occupancy, and so on.

(d) The physical condition and integrity of original construction and subsequent modifications are variables that must be considered in the analysis.

Furthermore, structures found in existing facilities were not only designed to lower levels of seismic demand but also with far fewer of the ductile detailing provisions found in today's codes, thus increasing the need for accurate assessment of these structures with appropriate levels of ductility.

4.6.2 Methods for Evaluation

An appropriate evaluation methodology should incorporate on-site visual assessment of the existing structures and systems, as described in detail in Chapter 6, in conjunction with appropriate analytical reviews or the use of tier-based evaluation criteria to identify and mitigate common seismic deficiencies in existing petrochemical facilities to reduce the level of seismic risk inherited in old structures to an acceptable limit defined by the owner. Two methods currently used to evaluate various structures in existing facilities are ASCE 41 (2013) and CalARP (2013).

ASCE 41 provides a three-tiered process for seismic evaluation of building structures in any level of seismicity. They are evaluated to either Life Safety or Immediate Occupancy Performance Level depending on their importance to the facility operation.

The three-tiered process is as follows:

(a) Screening Phase (Tier 1) consists of sets of checklists that allow a rapid evaluation of the structural, nonstructural, and foundation/geologic hazard elements of the building and site conditions. The purpose of Tier 1 is to quickly screen out buildings that are in compliance, or to identify potential deficiencies. The procedure used in this phase takes a different approach to accounting for the nonlinear seismic response. Pseudo-static lateral forces are applied to the structure to obtain "actual" displacements during a design earthquake, which does not represent the actual lateral force that the structure must resist in traditional design codes. In summary, this procedure is based on equivalent displacements and pseudo lateral forces. Also, instead of applying a single ductility-related response reduction factor (R) to the applied loads, a ductility-related m factor directly related to the ductility of each component of the structure is used in the acceptability check for that specific component.

(b) Evaluation Phase (Tier 2) provides detailed analyses of building deficiencies identified in Tier 1. The analysis procedure used in this phase includes the linear static procedure, the linear dynamic procedure, and special procedures using displacement-based lateral force and m factor on an element-by-element basis.

(c) Detailed Evaluation Phase (Tier 3) is necessary when the findings from Tier 1 and Tier 2 are determined to still be deficient and a more detailed evaluation would have a significant economic and other advantage. Tier 3 shall be performed using linear or nonlinear methods for static or dynamic analysis of the structure and comply with criteria and procedures described in ASCE 41.

CalARP, the California Accidental Release Prevention Program, was originally developed to satisfy the requirements of the state-mandated Risk Management and Prevention Program for facilities with threshold quantities of regulated substances (RS) as listed in California Code of Regulations Title 19 Division 2 Chapter 4.5. CalARP provides guidance regarding criteria to be used in seismically assessing structural systems and components whose failure would result in the release of sufficient quantities of RS to be of concern.

Those criteria differ from the traditional building code approach in the following areas:

(a) Loads are determined using a "Q" factor rather than the "R" factor used for new design. The Q factor is intended to address realistic conditions that will be found in existing facilities, including those configurations that would not be permitted in a new design.

(b) The seismic importance (Ie) factor is always set to unity (1.0) for evaluation of existing facilities unless it exceeds 1.0 for "special" structures at the owner's request. In contrast, for design of new facilities, the appropriate Ie factor, depending on the importance of the structure and functional requirements, may be higher than 1.0 (as much as 1.5).

(c) For evaluation of existing facilities, site-specific response spectra may be used without the consent of the building official. For new facilities, if a site-specific response spectrum is specified that differs from code-specified values, the consent of a building official must be obtained.

4.6.3 Overturning Potential

In evaluating existing structures, the safety factor against overturning should be limited to 1.0; however, a minimum of 10% reduction in dead load should be assumed to account for vertical acceleration effects. This reduction factor may be higher for facilities close to active faults. Alternatively, more refined overturning checks, such as the energy balance approach, may be used, and the check for vessels containing fluids should be completed in both full and dry conditions.

Effects of neighboring structures should also be considered. The energy balance approach checks that the potential energy to overturn the structure is not exceeded by the imposed kinetic energy caused by the earthquake. Strain energy effects in the structure and, more important, in the soil can be included.

Appendix 4.D further describes the energy balance approach for checking against overturning.

4.6.4 Sliding Potential

For structures in existing petrochemical facilities, a factor of safety of 1.0 against sliding is recommended for free-standing structures that are not anchored to their foundations, or structures resting directly on soil. The sliding check for vessels containing fluids should consider the vessel in both dry and full states. Effects on neighboring structures should also be considered.

In the event that sliding displacements are required (for example, for evaluation of tolerance requirements, pipe displacements, etc.), the methodology prescribed in Appendix 4.E can be used. The methodology considers the vertical upward accelerations present during earthquakes that can reduce the effective friction force between the structure and its support surface. The maximum sliding displacement can be related to the velocity imparted into the structure during the earthquake.

Time history studies of dynamic stability show that the calculated maximum inertial forces act for only short periods of time, reversing in direction many times during an earthquake. Considering these maximum forces as static forces for purposes of stability and sliding analysis is very conservative. In fact, the factor of safety against sliding and overturning has been shown to drop below unity for short periods of time during the seismic event without failure. This is because any rigid body movement is arrested when dynamic forces decrease or reverse directions.

References

ACI (American Concrete Institute). 2014. *Building code requirements for structural concrete.* ACI 318. Farmington Hills, MI: ACI.

AISC. 2016a. *Seismic provisions for structural steel buildings.* AISI 341. Chicago: AISC.

AISC. 2016b. *Specification for structural steel buildings.* AISC 360. Chicago: AISC.

API (American Petroleum Institute). 2009. *Management of hazards associated with location of process plant permanent buildings.* API RP 752. Washington, DC: API.

API. 2014a. *Design and construction of large, welded, low pressure storage tanks,* 12th ed. Washington, DC: API.

API. 2014b. *Welded steel tanks for oil storage,* 12th ed. Washington, DC: API.

API. 2016. *Fired heaters for general refinery service.* API 560. Washington, DC: API.

ASCE. 2013. *Seismic evaluation and retrofit of existing buildings.* ASCE 41. Reston, VA: ASCE.

ASCE. 2016. *Minimum design loads and associated criteria for buildings and other structures.* ASCE/SEI 7-16. Reston, VA: ASCE.

AWWA (American Water Works Association). 2009. *Factory-coated bolted steel tanks for water storage.* AWWA D103. Denver: AWWA.

AWWA. 2011. *Welded steel tanks for water storage.* AWWA D100. Denver: AWWA.

Bechtel Power. 1980. "Seismic analyses of structures and equipment for nuclear power plants." In *Bechtel design guide C2.44, Rev. 0.* Reston, VA: Bechtel Power.

Bowels, J. E. 1996. *Foundation analysis and design,* 5th ed. New York: McGraw-Hill.

CalARP. 2013. *Guidance for California accidental release prevention (CalARP) program seismic assessments.* Sacramento, CA: CalARP Program Seismic Guidance Committee.

Caltrans (State of California Department of Transportation). 2015. "Seismic analysis of bridge structures." In *Caltrans bridge design practice,* 4th ed. Sacramento, CA: Caltrans.

Esfandiari, S., and P. B. Summers. 1994. "Seismic analysis and load determination in support of seismic evaluation and design of petrochemical facilities." In *Proc., American Power Conf.,* Chicago.

FEMA. 2012. *Reducing the risks of nonstructural earthquake damage: A practical guide.* FEMA E-74. Washington, DC: FEMA.

IBC (International Building Code). 2018. *International building code.* Country Club Hills, IL: International Code Council.

Newmark, N. W., and W. J. Hall. 1982. *Earthquake spectra and design.* Oakland, CA: Earthquake Engineering Research Institute.

Wolf, J. P. 1985. *Dynamic soil-structure interaction.* Englewood Cliffs, NJ: Prentice-Hall.

APPENDIX 4.A TYPICAL PERIOD (T) COMPUTATIONS FOR NONBUILDING STRUCTURES

NOTE: Equations in this appendix are presented in US customary units only, as coefficients are developed for US customary units only.

A. METHODOLOGY

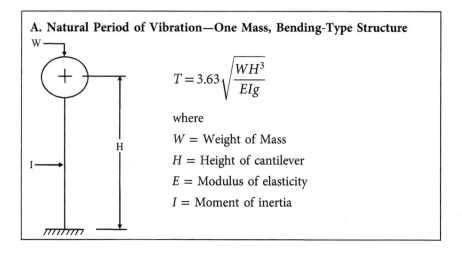

A. Natural Period of Vibration—One Mass, Bending-Type Structure

$$T = 3.63 \sqrt{\frac{WH^3}{EIg}}$$

where

W = Weight of Mass

H = Height of cantilever

E = Modulus of elasticity

I = Moment of inertia

B. Natural Period of Vibration—One Mass, Rigid Frame-Type Structure

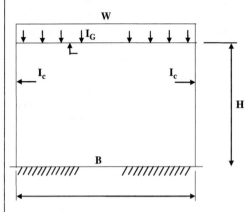

$$T = 1.814 \sqrt{\frac{\alpha W H^3}{EI_c g}}$$

For Columns Hinged at Base

$$\alpha = \frac{2K + 1}{K}$$

For Columns Fixed at Base

$$\alpha = \frac{3K + 2}{6K + 1}$$

$$K = \left(\frac{I_G}{I_c}\right)\left(\frac{H}{B}\right)$$

C. Natural Period of Vibration—Two Mass Structure

$$T = 2\pi \sqrt{\frac{W_A C_{aa} + W_B C_{bb} + \sqrt{(W_A C_{aa} - W_B C_{bb})^2 + 4 W_A W_B C_{ab}^2}}{2g}}$$

where

C_{aa} = Deflection at a due to unit lateral load at a

C_{bb} = Deflection at B due to unit lateral load at B

C_{ab} = Deflection at B due to unit lateral load at A

W_A, W_B = Summation of vertical loads at level a or b

See Example 1 for application.

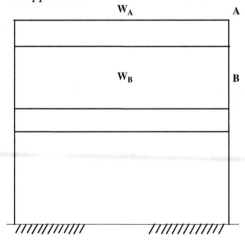

D. Natural Period of Vibration—Bending-Type Structure, Uniform Weight Distribution and Constant Cross Section

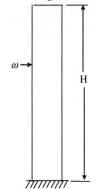

$$T = 1.79\sqrt{\frac{\omega H^4}{EIg}}$$

where
ω = weight per unit height

E. Natural Period of Vibration—Uniform Vertical Cylindrical Steel Vessel

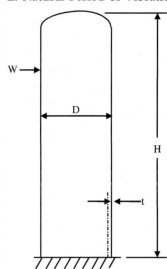

$$T = \frac{7.78}{10^6}\left(\frac{H}{D}\right)^2\sqrt{\frac{12WD}{t}}$$

where

T = Period (s)

W = Weight (lb/ft)

H = Height (ft)

D = Diameter (ft)

t = Shell thickness (in.)

F. Natural Period of Vibration—Nonuniform Vertical Cylindrical Vessel

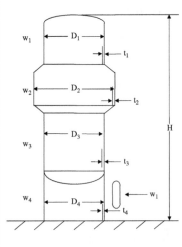

$$T = \left(\frac{H}{100}\right)^2 \cdot \sqrt{\frac{\sum w.\Delta\alpha + \frac{1}{H} \cdot \sum W.\beta}{\sum E.D^3.t.\Delta\gamma}}$$

where

T = Period (s)

H = Overall height (ft)

w = Distributed weight (lb/ft) of each section

W = Weight (lb) of each concentrated mass

D = Diameter (ft) of each section

t = Shell thickness (in.) of each section

E = Modulus of elasticity (millions of psi) α, β, and γ are coefficients for a given level depending on h_x/H ratio of the height of the level above grade to the overall height. $\Delta\alpha$ and $\Delta\gamma$ are the difference in the values of α and γ, from the top to the bottom of each section of uniform weight, diameter, and thickness. β is determined for each concentrated mass. Values of α, β, and γ are tabulated on the following table.

Coefficients for Determining Period of Vibration of Free-Standing Cylindrical Shells with Nonuniform Cross Section and Mass Distribution

h_x/H	α	β	γ	h_x/H	α	β	γ
1.00	2.103	8.347	1.000000	0.50	0.1094	0.9863	0.95573
0.99	2.021	8.121	1.000000	0.49	0.0998	0.9210	0.95143
0.98	1.941	7.898	1.000000	0.48	0.0909	0.8584	0.94683
0.97	1.863	7.678	1.000000	0.47	0.0826	0.7987	0.94189
0.96	1.787	7.461	1.000000	0.46	0.0749	0.7418	0.93661

(Continued)

Coefficients for Determining Period of Vibration of Free-Standing Cylindrical Shells with Nonuniform Cross Section and Mass Distribution (Continued)

h_x/H	α	β	γ	h_x/H	α	β	γ
0.95	1.714	7.248	0.999999	0.45	0.0578	0.6876	0.93097
0.94	1.642	7.037	0.999998	0.44	0.0612	0.6361	0.92495
0.93	1.573	6.830	0.999997	0.43	0.0551	0.5372	0.91854
0.92	1.506	6.626	0.999994	0.42	0.0494	0.5409	0.91173
0.91	1.440	6.425	0.999989	0.41	0.0442	0.4971	0.90443
0.90	1.377	6.227	0.999982	0.40	0.0395	0.4557	0.89679
0.89	1.316	6.032	0.999971	0.39	0.0351	0.4167	0.88864
0.88	1.256	5.840	0.999956	0.38	0.0311	0.3801	0.88001
0.87	1.199	5.652	0.999934	0.37	0.0275	0.3456	0.87033
0.86	1.143	5.467	0.999905	0.36	0.0242	0.3134	0.86123
0.85	1.090	5.285	0.999867	0.35	0.0212	0.2833	0.85105
0.84	1.038	5.106	0.999817	0.34	0.0185	0.2552	0. 84032
0.83	0.938	4.930	0.999754	0.33	0.0161	0.2291	0.82901
0.82	0.939	4.758	0.999674	0.32	0.0140	0.2050	0.81710
0.81	0.892	4.589	0.999576	0.31	0.0120	0.1826	0.80459
0.80	0.847	4.424	0.999455	0.30	0.010293	0.16200	0.7914
0.79	0.804	4.261	0.999309	0.29	0.008769	0.14308	0.7776
0.78	0.762	4.102	0.999133	0.28	0.007426	0.12576	0.7632
0.77	0.722	3.946	0.998923	0.27	0.006249	0.10997	0.7480
0.76	0.683	3.794	0.998676	0.26	0.005222	0.09564	0.7321
0.75	0.646	3.645	0.998385	0.25	0.004332	0.08267	0.7155
0.74	0.610	3.499	0.998047	0.24	0.003564	0.07101	0.6981
0.73	0.576	3.356	0.997656	0.23	0.002907	0.06056	0.6800
0.72	0.543	3.217	0.997205	0.22	0.002349	0.05126	0.6610
0.71	0.512	3.081	0.996689	0.21	0.001878	0.04303	0.6413
0.70	0.481	2.949	0.996101	0.20	0.001485	0.03579	0.6207
0.69	0.453	2.820	0.995434	0.19	0.001159	0.02948	0.5902
0.68	0.425	2.694	0.994681	0.18	0.000893	0.02400	0.5769
0.67	0.399	2.571	0.993834	0.17	0.000677	0.01931	0.5536
0.66	0.374	2.452	0.992885	0.16	0.000504	0.01531	0.5295
0.65	0.3497	2.3365	0.99183	0.15	0.000368	0.01196	0.5044
0.64	0.3269	2.2240	0.99065	0.14	0.000263	0.00917	0.4783
0.63	0.3052	2.1148	0.98934	0.13	0.000183	0.00689	0.4512
0.62	0.2846	2.0089	0.98739	0.12	0.000124	0.00506	0.4231
0.61	0.2650	1.9062	0.98630	0.11	0.000081	0.00361	0.3940

(Continued)

Coefficients for Determining Period of Vibration of Free-Standing Cylindrical Shells with Nonuniform Cross Section and Mass Distribution (Continued)

h_x/H	α	β	γ	h_x/H	α	β	γ
0.60	0.2464	1.8068	0.98455	0.10	0.000051	0.00249	0.3639
0.59	0.2288	1.7107	0.98262	0.09	0.000030	0.00165	0.3327
0.58	0.2122	1.6177	0.98052	0.08	0.000017	0.00104	0.3003
0.57	0.1965	1.5279	0.97823	0.07	0.000009	0.00062	0.2669
0.56	0.1816	1.4413	0.97573	0.06	0.000004	0.00034	0.2323
0.55	0.1676	1.3579	0.97301	0.05	0.000002	0.00016	0.1965
0.54	1.1545	1.2775	0.97007	0.04	0.000001	0.00007	0.1597
0.53	0.1421	1.2002	0.96683	0.03	0.000000	0.00002	0.1216
0.52	0.1305	1.1259	0.96344	0.02	0.000000	0.00000	0.0823
0.51	0.1196	1.0547	0.95973	0.01	0.000000	0.00000	0.0418

G. Period of Vibration—Generalized One Mass Structure

$$T = 2\pi \left(\frac{y}{g}\right)^{0.5}$$

where

y = Static deflection of mass resulting from a lateral load applied at the mass equal to its own weight.

g = Acceleration due to gravity

See Example 4 for application.

H. Impulsive Period of Vibration—Flat Bottom Liquid Storage Tank

$$T_i = \frac{C_i H}{27.8 \sqrt{\frac{t_u}{D}}} \left[\frac{\sqrt{\gamma}}{\sqrt{E}}\right]$$

where

C_i = Coefficient for determining impulsive period of tank system read from the following figure

H = Maximum design product level, ft

γ = Density of stored liquid, lb/ft^3

t_u = Equivalent uniform thickness of tank shell, in.

D = Nominal tank diameter, ft

E = Elastic modulus of tank material, psi

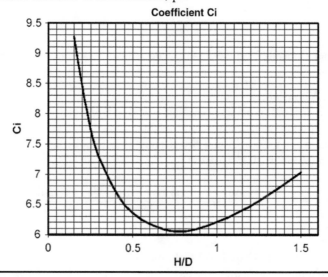

Coefficient Ci

I. Convective Period of Vibration—Flat Bottom Liquid Storage Tank

$$T_c = 2\pi \sqrt{\frac{D}{3.68\, g\, \tanh\left(\frac{3.68\, H}{D}\right)}}$$

where

H = Maximum design product level, ft

g = Acceleration due to gravity = 32.2 ft/s^2

D = Nominal tank diameter, ft

B. EXAMPLES

Example 1. Two-Story Concrete Vessel Support Structure

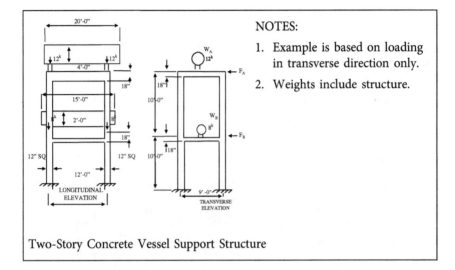

NOTES:

1. Example is based on loading in transverse direction only.

2. Weights include structure.

Two-Story Concrete Vessel Support Structure

Period of Vibration:
Earthquake forces (Transverse direction—Loads on one bent)
   $W = 20$ kips (includes structure weight)
Deflections from 1 kip at A and B (calculations not shown):
   $C_{aa} = 0.0384$ in., $C_{ab} = 0.0180$ in., $C_{bb} = 0.0157$ in.

$$T = 2\,(3.14)\left\{\frac{12\,(0.0384) + 8\,(0.0157) + [[12\,(0.0384) - 8\,(0.0157)]^2 + 4\,(12)\,(8)\,(0.0180)^2]^{0.5}}{2\,(386)}\right\}^{0.5}$$

$$= 0.234\,\text{s}$$

Example 2. Uniform Cylindrical Column
In most columns of constant diameter, the entire mass can be assumed uniformly distributed over the height. Where concentrations of mass are large or variations in cross-section occur, the analysis should be made as shown in Example 3.

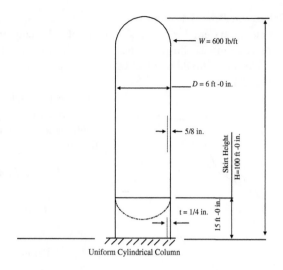

Uniform Cylindrical Column

Period of Vibration:

$$T = \frac{7.78}{10^6} \left(\frac{100}{6}\right)^2 \left(\frac{12 \times 600 \times 6}{0.25}\right)^{0.5} = 0.898 \text{ s}$$

Example 3. Column with Variable Cross Section and Mass Distribution

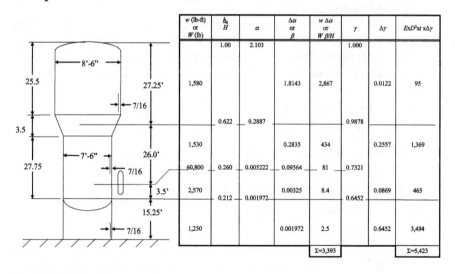

w (lb-ft) or W (lb)	$\frac{h_g}{H}$	α	$\Delta\alpha$ or β	$w\,\Delta\alpha$ or $W\,\beta/H$	γ	$\Delta\gamma$	$E \times D^3 \times t \times \Delta\gamma$
	1.00	2.103			1.000		
1,580			1.8143	2,867		0.0122	95
	0.622	0.2887			0.9878		
1,530			0.2835	434		0.2557	1,369
60,800	0.260	0.005222	0.09564	81	0.7321		
2,570			0.00325	8.4		0.0869	465
	0.212	0.001972			0.6452		
1,250			0.001972	2.5		0.6452	3,494
				Σ=3,393			Σ=5,423

Period of Vibration:

$$T = (72/100)^2 (3{,}393/5{,}423)^{0.5} = 0.41 \text{ s}$$

Example 4. Sphere on Braced Columns

The common bracing system for spheres consists of x-bracing that connects adjacent pairs of columns as illustrated below. The bracing for large spheres subject to earthquake loads should be effective both in tension and compression to better resist the lateral forces.

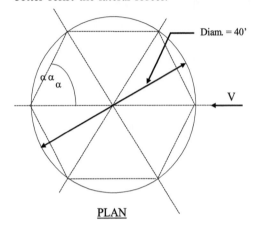

Diam. = 40'

V

PLAN

The shear in each panel and the max maximum panel shear could be found by the following formula:

$$V_p = (2V/n)\cos\alpha$$

$$V_{pmax} = 2V/n$$

where

V_p = Panel shear

V_{pmax} = Maximum panel shear

V = Lateral force

n = Number of panels

α = Angle between plane of panel and direction of lateral force

Period of Vibration:

The period of vibration is found using the general formula for one-mass structure previously. The static deflection, y, is found by determining the change in length of the bracing resulting from a total lateral load equal to the weight of the sphere. Deformation of the columns and balcony girder are usually neglected for one story.

P = Maximum force in brace

= (1/2) (2×1,500/6) (36.0/20.0) = 450 kips

Δ = Change in length of brace

= PL/EA = [(450) (36.0) (12)]/[(29,000 (8.0)] = 0.838 in.

Y = Δ/sin θ = (0.838) (36.0/20) = 1.51 in.

Period of Vibration, $T = 2\pi [y/g]^{0.5}$

$T = 2\pi [1.51/(32.2)(12)]^{0.5} = 0.393$

Note: If the sphere used tension-only bracing, the period T would be 0.556 s.

Example 5. Ground-Supported Liquid Storage Tank
Determine the impulsive and convective periods for a 64 ft-0 diameter by 32 ft-0 high water tank shown in the figure below.

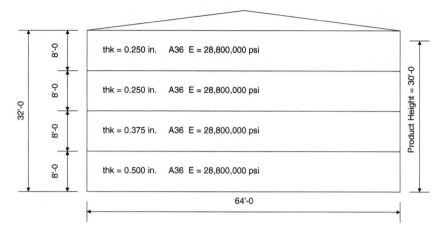

Impulsive Period

    $D = 64$ ft $H = 30$ ft

    $H/D = 0.47$ $C_i = 6.4$ (from figure in Section H)

    The equivalent uniform thickness of the tank wall, t_u, is calculated by the weighted average method. Using weights equal to the distance from the liquid surface:

$$t_u = \frac{0.25\,(8)\left(\frac{6}{2}\right) + 0.25\,(8)\left(6 + \frac{8}{2}\right) + 0.375\,(8)\left(6 + 8 + \frac{8}{2}\right) + 0.5\,(8)\left(6 + 8 + 8 + \frac{8}{2}\right)}{(8)\left(\frac{6}{2}\right) + (8)\left(6 + \frac{8}{2}\right) + (8)\left(6 + 8 + \frac{8}{2}\right) + (8)\left(6 + 8 + 8 + \frac{8}{2}\right)} = 0.4035$$

$$\gamma = 62.4\,\text{lb}/\text{ft}^3$$

$$T_i = \frac{C_i H}{27.8\sqrt{\frac{t_u}{D}}}\left[\frac{\sqrt{\gamma}}{\sqrt{E}}\right] = \frac{6.4\,(30)}{27.8\sqrt{\frac{0.4035}{64}}}\left(\sqrt{\frac{62.4}{28{,}800{,}000}}\right) = 0.128\,\text{s}$$

Convective Period

$$T_c = 2\pi\sqrt{\frac{D}{3.68g\,\tanh\left(\frac{3.68H}{D}\right)}} = 2\pi\sqrt{\frac{64}{3.68\,(32.2)\,\tanh\left(\frac{3.68\,(30)}{64}\right)}} = 4.767\,\text{s}$$

APPENDIX 4.B GUIDELINES FOR DETERMINING BASE SHEAR FOR COMBINATION STRUCTURES

A. METHODOLOGY

NONBUILDING STRUCTURES SUPPORTED ABOVE GRADE
$$W_p \le 0.25 \ (W_S + W_p)$$

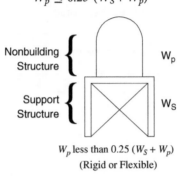

Nonbuilding Structure $\left\{ \right.$ W_p

Support Structure $\left\{ \right.$ W_S

W_p less than 0.25 $(W_S + W_p)$
(Rigid or Flexible)

Support structure similar to building (lump mass of item with mass of structure):

$$V = C_S(W_p + W_S)$$

$$C_S = \text{lesser of the following:} \ \frac{S_{DS}}{\left(\frac{R}{I_e}\right)} \ \text{or} \ \frac{S_{D1}}{T\left(\frac{R}{I_e}\right)} \ \text{for T} \le T_L \ \text{or} \ \frac{S_{D1}T_L}{T^2\left(\frac{R}{I_e}\right)} \text{for T} > T_L$$

but not less than $0.044 S_{DS} \ I_e \ge 0.01$ or $\frac{0.5 S_1}{\left(\frac{R}{I_e}\right)}$ where $S_1 \ge 0.6$ g

Nonbuilding structure (NBS) and anchorage:

$$F_p = \frac{0.4 \, a_p \, S_{DS} \, W_p}{\left(\frac{R_p}{I_p}\right)} \left(1 + 2\frac{z}{h}\right) = \frac{1.2 \, a_p \, S_{DS} \, W_p}{\left(\frac{R_p}{I_p}\right)} \qquad \text{Applied at centroid on NBS.}$$

but not less than $0.3 \ S_{DS} \ I_p \ W_p$ or greater than $1.6 \ S_{DS} \ I_p \ W$

CASE 1

Note: Values of R for nonbuilding structures similar to buildings are in ASCE 7 Table 12.2-1 or Table 15.4-1, while values of ap and Rp are in Table 13.6-1. See Section 4.4.3 for additional discussion on the correct value of a_p to use.

NONBUILDING STRUCTURES SUPPORTED ABOVE GRADE
$$W_p > 0.25 \ (W_S + W_p)$$

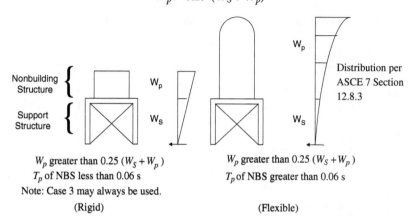

Distribution per ASCE 7 Section 12.8.3

W_p greater than 0.25 $(W_S + W_p)$
T_p of NBS less than 0.06 s
Note: Case 3 may always be used.
(Rigid)

W_p greater than 0.25 $(W_S + W_p)$
T_p of NBS greater than 0.06 s
(Flexible)

Support structure:

$V = C_S(W_p + W_S)$
$C_S = $ lesser of the following: $\frac{S_{DS}}{\left(\frac{R}{I_e}\right)}$

or $\frac{S_{D1}}{T\left(\frac{R}{I_e}\right)}$ for $T \leq T_L$

or $\frac{S_{D1} T_L}{T^2\left(\frac{R}{I_e}\right)}$ for $T > T_L$

but not less than $0.044\, S_{DS}\, I_e \geq 0.03$
or $\frac{0.8 S_1}{\left(\frac{R}{I_e}\right)}$ where $S_1 \geq 0.6$ g

(Lump mass of item with mass of structure)

Nonbuilding structure and anchorage:

$V_1 = F_p = \frac{0.4\, a_p\, S_{DS}\, W_p}{\left(\frac{R_p}{I_p}\right)}\left(1 + 2\frac{z}{h}\right) = \frac{1.2\, a_p\, S_{DS}\, W_p}{\left(\frac{R_p}{I_p}\right)}$

but not less than $0.3\, S_{DS}\, I_p\, W_p$
or greater than $1.6\, S_{DS}\, I_p\, W_p$
$R_p = R$ of nonbuilding structure
and $a_p = 1$

Support structure and nonbuilding structure:

$V = C_S(W_p + W_S)$
$C_S = $ lesser of the following: $\frac{S_{DS}}{\left(\frac{R}{I_e}\right)}$

or $\frac{S_{D1}}{T\left(\frac{R}{I_e}\right)}$ for $T \leq T_L$

or $\frac{S_{D1} T_L}{T^2\left(\frac{R}{I_e}\right)}$ for $T > T_L$

but not less than
$0.044\, S_{DS}\, I_e \geq 0.03$
or $\frac{0.8 S_1}{\left(\frac{R}{I_e}\right)}$ where $S_1 \geq 0.6$ g

R is lesser value of supporting structure or nonbuilding structure.

Applied at centroid of NBS.

CASE 2   CASE 3

Note: Values of R for nonbuilding structures similar to buildings are in ASCE 7
Table 12.2-1 or Table 15.4-1, while values of ap and Rp are in Table 13.6-1.
R values for nonbuilding structures not similar to buildings are in ASCE 7
Table 15.4-2.

B. EXAMPLES

Example: Table-Top Reinforced Concrete Structure Supporting Reactor Vessels

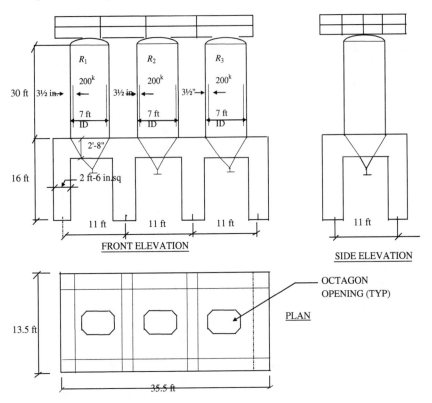

FRONT ELEVATION

SIDE ELEVATION

PLAN

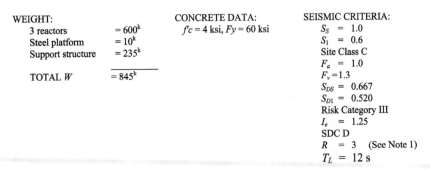

DESIGN DATA:

WEIGHT:		CONCRETE DATA:	SEISMIC CRITERIA:	
3 reactors	$= 600^k$	$f'c = 4$ ksi, $Fy = 60$ ksi	S_S	$= 1.0$
Steel platform	$= 10^k$		S_1	$= 0.6$
Support structure	$= 235^k$		Site Class C	
			F_a	$= 1.0$
TOTAL W	$= 845^k$		$F_v = 1.3$	
			S_{DS}	$= 0.667$
			S_{D1}	$= 0.520$
			Risk Category III	
			I_e	$= 1.25$
			SDC D	
			R	$= 3$ (See Note 1)
			T_L	$= 12$ s

DETERMINE STRUCTURAL PERIOD, T :

By inspection, each of the reactors is rigid and thus the fundamental period of the combined structure will be determined based on the stiffness of the supporting structure. If the reactor vessels are not rigid (i.e., $T > 0.06$ s), then the structure may be considered as a two lumped mass system whose period may be determined using other methods.

1. TRANSVERSE DIRECTION

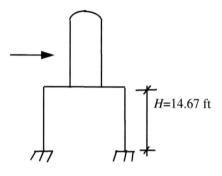

$H = 14.67$ ft

Assume the table top slab is a rigid diaphragm (See Note 2):

Support stiffness per Bent :

For simplicity, assume as a moment with infinitely rigid girder.

$$K = \frac{12EI}{H^3} \times 2$$

$$= \frac{12 \times 3605 \times 33750 \times 2}{(14.67 \times 12)^3}$$

$$= 535 \text{ kips/in. (per bent)}$$

Member properties (See Note 3):

$$\Sigma K = 4 \times 535 = 2140 \text{ k/in.}$$

$$(I_C)_{gross} = \frac{bh^3}{12} = \frac{(30)(30)^3}{12} = 67{,}500 \text{ in.}^4$$

Assume $(I_C)\,eff = 50\% \times I_{gross} = 33{,}750 \text{ in.}^4$

$$E_C = 57\sqrt{4000} = 3605 \text{ ksi}$$

Fundamental period:

$$T = 2\pi\sqrt{\frac{M}{\Sigma K}}$$

where:

M = total mass of vessel + support

K = support stiffness

$$T = 2\pi\sqrt{\frac{845}{2140 \times 386.4}} = 0.20 \text{ Sec.}$$

Note 1:   The supporting structure is a concrete intermediate moment frame. To use this system in SDC D, a reduced R value of 3 is chosen corresponding to "With permitted height increase" from ASCE 7 Table 15.4-1.

Note 2:   If the diaphragm is not rigid, then the period calculation should be performed based on the tributary mass per bent.

Note 3:   The effective moment of inertia of columns can be approximated as 50% of the gross moment of inertia (Chapter 8 "Seismic Analysis of Bridge Structures" from Caltrans Bridge Design Practice).

2. LONGITUDINAL DIRECTION

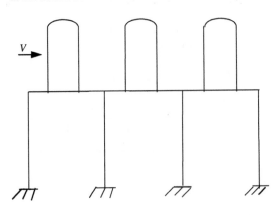

As in the transverse direction, the structural period of the support can be determined as follows:

$$K = \frac{12EI}{H^3} \times 4$$

$$= \frac{12 \times 3605 \times 33750 \times 4}{(14.67 \times 12)^3}$$

$$= 1070\,k/in\,(per\,bent)$$

$$\Sigma K = 2 \times 1070 = 2140\,k/in.$$

Fundamental period:

$$T = 2\pi\sqrt{\frac{M}{\Sigma K}}$$

$$= 2\pi\sqrt{\frac{845}{2141 \times 386.4}} = 0.2\,Sec$$

DETERMINE BASE SHEAR:

Because the period and R are both the same in both directions, so too will be the base shear:

$$V = C_S(W_p + W_S)$$

C_S = lesser of the following: $\frac{S_{DS}}{(\frac{R}{I_e})} = \frac{0.667}{(\frac{3}{1.25})} = 0.278(governs)$

or $\frac{S_{D1}}{T(\frac{R}{I_e})} = \frac{0.52}{0.2(\frac{3}{1.25})} = 1.083$ for $T \le T_L$

but not less than 0.03 or $\dfrac{0.8S_1}{\left(\frac{R}{I_e}\right)} = \dfrac{0.8(0.6)}{\left(\frac{3}{1.25}\right)} = 0.20$ where $S_1 \geq 0.6$ g

$$V = C_S\,(W_p + W_S) = C_S\,(845)$$
$$= 0.278\,(845)$$
$$= 235 \text{ kips (strength level)}$$

LATERAL FORCE DISTRIBUTION:

Because the subject table top structure has a rigid diaphragm, the lateral force is distributed to each bent per its relative stiffness. Because the stiffness of each bent is the same, the lateral force to each bent is essentially identical.

TRANSVERSE DIRECTION:

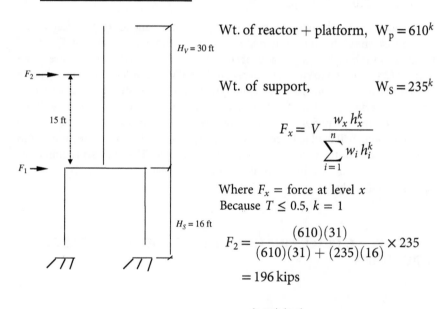

Wt. of reactor + platform, $W_p = 610^k$

Wt. of support, $\qquad W_S = 235^k$

$$F_x = V\,\dfrac{w_x\,h_x^k}{\displaystyle\sum_{i=1}^{n} w_i\,h_i^k}$$

Where F_x = force at level x
Because $T \leq 0.5,\ k = 1$

$$F_2 = \dfrac{(610)(31)}{(610)(31) + (235)(16)} \times 235$$
$$= 196 \text{ kips}$$

$$F_1 = \dfrac{(235)(16)}{(610)(31) + (235)(16)} \times 235 = 39 \text{ kips}$$

LONGITUDINAL DIRECTION: Same as for transverse direction calculated above.

NONBUILDING STRUCTURE (REACTOR) AND ANCHORAGE:

The reactor and its connections to the concrete frame are to be designed for the following force:

$$F_p = \frac{0.4\, a_p\, S_{DS}\, W_p}{\left(\frac{R_p}{I_p}\right)} \left(1 + 2\frac{z}{h}\right)$$

$R_p = R$ of nonbuilding structure $= 2$

$$= \frac{1.2\, a_p\, S_{DS}\, W_p}{\left(\frac{R_p}{I_p}\right)}$$

$a_p = 1$ and $I_p = 1.5$

but not less than $03\, S_{DS}\, I_p\, W_p$ or greater than $1.6\, S_{DS}\, I_p\, W_p$

$$F_p = \frac{1.2\,(1)\,(0.667)\,(610)}{\left(\frac{2}{1.5}\right)} = 366 \text{ kips (strength level)}$$ (governs – applied at centroid of NBS)

$1.6\, S_{DS}\, I_p\, W_p = 1.6(0.667)(1.5)(610) = 976$ kips and $0.3\, S_{DS}\, I_p\, W_p = 0.3$ $(0.667)(1.5)(610) = 813$ kips

Please note that F_p (the design force for the NBS and its anchorage) is greater than V (the base shear) for the combined structure. The increased value of F_p over V results from the higher importance factor (1.5 vs. 1.25) used and the lower NBS R-value (2 vs. 3) used to calculate forces on the NBS. The difference in importance factors between a nonstructural component and other structures covered by ASCE 7 is a disconnect in ASCE 7-16 and will be corrected in a future edition. Because of the simplifying assumptions used in the procedure (Case 2), the resulting forces will tend to be conservative. Case 3, described under Section A, can be used instead of the procedure used in this example. The analysis required for Case 3 may result in lower design forces, but will require a much more sophisticated analysis.

APPENDIX 4.C DETERMINATION OF BASE SHEAR FOR SELECTED STRUCTURES

Table 4.C-1. Determination of Base Shear for Selected Structures.

Note: Structure period, T, assumed to be $\geq T_O$ and $\leq T_S = S_{D1}/S_{DS}$ for all examples in this table.

Equipment/Structure Types	Components	R	Base Shear Equation			Remarks
1.0 Vertical vessels			$V = \dfrac{S_{DS}W}{R/I_e}$	$F_p = \dfrac{0.4a_p S_{DS} W_p (1+2z/h)}{R_p/I_p}$	$V = 0.35 S_{DS} I_e W$	
1.1 Supported on skirt	Vessel, skirt, and anchorage	2.0	$0.5 S_{DS} I_e W$			If structure is in Risk Category IV, the vessel and skirt must be checked for seismic loads determined using $I_e/R = 1$ with $FS = 1$. $R = 3$ may be used if the vessel and skirt are checked for seismic loads determined using $I_e/R = 1$ with $FS = 1$.
		3.0	$0.33 S_{DS} I_e W$			

(Continued)

Table 4.C-1. Determination of Base Shear for Selected Structures. (Continued)

Equipment/Structure Types	Components	R	Base Shear Equation	Remarks
1.0 Vertical vessels		$V = \dfrac{S_{DS}W}{R/I_e}$	$F_p = \dfrac{0.4 a_p S_{DS} W_p (1+2z/h)}{R_p/I_p}$	$V = 0.3 S_{DS} I_e W$
1.2 Supported on structure (a) For $W_P \le 25\%$ $(W_S + W_P)$ steel/concrete, special moment frame (SMRF)	Vessel and anchorage	See	$1.2 S_{DS} I_p W_p$	Skirt-supported pressure vessel $a_p = 2.5$ and $R_p = 2.5$ $z = h$
	Support structure	Table 4.C-2		Nonbuilding structure with structural system similar to building
Steel intermediate moment frame (IMRF)				
Steel ordinary moment frame (OMRF)				
Concrete Intermediate Moment Frame (IMRF)				
Concrete Ordinary Moment Frame (OMRF)				
(b) For $W_P > 25\%$ $(W_S + W_P)$ Steel and conc. SMRF, steel OMRF, and steel and conc. IMRF	Vessel and anchorage	2.0	$0.6 S_{DS} I_p W_p$	Rigid vessel $a_p = 1$ and $z=h$ Flexible vessel Use min. R of vessel or frame
	Support structure	Same as 1.2.a	$\dfrac{S_{DS} I_e W}{R}$	
	Support structure, vessel, and anchorage			

Equipment/Structure Types	Components	R	Base Shear Equation	Remarks
			$V = \frac{S_{DS}W}{R/I_e}$ $F_p = \frac{0.4a_p S_{DS}W_p(1+2z/h)}{R_p/I_p}$ $V=0.3S_{DS}I_eW$	
2.0 Horizontal vessels/exchangers				
2.1 On short/rigid piers				
For both longitudinal and transverse directions ($T < 0.06$ s)	Vessel and anchorage	3.0	$0.33S_{DS}I_eW$	
	Support structure		$0.3S_{DS}I_eW$	Rigid nonbuilding structure
2.2 On flexible piers/tees				
For both longitudinal and transverse directions $T > 0.06$ s	Vessel and anchorage		$0.48S_{DS}I_pW_p$	$a_p = 1.0$ $R_p = 2.5$ $h=z$ Considered as inverted pendulum structure
	Support structure	2.0	$0.50S_{DS}I_eW$	

Table 4.C-1. Determination of Base Shear for Selected Structures. (Continued)

Equipment/Structure Types	Components	R	Base Shear Equation			Remarks
			$V = \frac{S_{DS}W}{R/I_e}$	$F_p = \frac{0.4a_pS_{DS}W_p(1+2z/h)}{R_p/I_p}$	$V = 0.35_{DS}I_eW$	
3.0 Aircooled exchangers (fin fans)						
3.1 Mounted at grade						
(a) Longitudinal bent concentric steel brace frame (OCBF)		See Table 4.C-2				Nonbuilding structure with a structural system similar to buildings
(b) Transverse bent steel OMRF		See Table 4.C-2				Same as previous
3.2 Mounted on top of pipeway (less than 25% of combined weight of NBS and support structure)						
(a) Transverse bent	Finfan design			$1.0S_{DS}I_pW_p$		$a_p = 2.5$ $R_p = 3.0$
	Pipeway design	See Table 4.C-2				Include reaction from fin fan
(b) Longitudinal bent	Fin fan design			$1.0S_{DS}I_pW_p$		$a_p = 2.5$ $R_p = 3.0$
	Pipeway design	See Table 4.C-2				Include reaction from fin fan
	Pipeway design	See Table 4.C-2				Same as previous

Steel OMRF

Steel and conc. SMRF, IMRF, or OMRF

Steel concentric braced frame (OCBF)

Steel concentric braced frame (OCBF)

Concrete SMRF, IMRF, or OMRF

Equipment/Structure Types	Components	R	Base Shear Equation $V = \frac{S_{DS}W}{R/I_e}$ $\quad F_p = \frac{0.4a_p S_{DS}W_p(1+2z/h)}{R_p/I_p}$ $\quad V = 0.35_{DS}I_eW$	Remarks
4.0 Pipeway Transverse / Longitudinal				
4.1 Steel frame main pipeway				
(a) Transverse bent SMRF, IMRF, OMRF, EBF, OCBF		See Table 4.C-2		Nonbuilding structure with structural system similar to building
(b) Longitudinal bent		See Table 4.C-2		
4.2 Concrete frame main pipeway				
(a) Transverse bent SMRF, IMRF, OMRF		See Table 4.C-2		Nonbuilding structure with structural system similar to building
(b) Longitudinal bent SMRF, IMRF, OMRF		See Table 4.C-2		
4.3 Steel/concrete cantilever stanchion	Inverted pendulum	2.0	$0.50S_{DS}I_eW$	Note difference between pipeway and inverted pendulum

Table 4.C-1. Determination of Base Shear for Selected Structures. (Continued)

Equipment/Structure Types	Components	R	Base Shear Equation			Remarks
			$V = \frac{S_{DS}W}{R/I_e}$	$F_p = \frac{0.4a_pS_{DS}W_p(1+2z/h)}{R_p/I_p}$	$V=0.3S_{DS}I_eW$	
5.0 Horizontal box heater/furnace						
5.1 Longitudinal braced frame and/or shear panel	Steel OCBF or shear panel design (usually redundant system)	See	Table 4.C-2			Nonbuilding structure with structural system similar to building
5.2 Transverse moment frame	Steel OMRF	See	Table 4.C-2			Nonbuilding structure with structural system similar to building
5.3 Tall concrete pier/pedestal	Pier/pedestal	2.0	$0.50S_{DS}I_eW$			Inverted pendulum type system

Equipment/Structure Types	Components	R	Base Shear Equation $V = \frac{S_{DS}W}{R/I_e}$	$F_p = \frac{0.4a_p S_{DS} W_p (1+2z/h)}{R_p/I_p}$	$V=0.3S_{DS}I_eW$	Remarks
6.0 Vessel on braced/unbraced legs						
6.1 Sphere (without top girder or stiffening ring)	Sphere and braced frame design	3.0	$0.33S_{DS}I_eW$			Nonbuilding structure
						Use same base shear for design of anchorage and support structure. Also see ASCE 7 Section 15.7.10.4d.
6.2 Stack/cylindrical furnace, vertical vessel, or hopper on cantilever or braced legs (without top girder or stiffening ring)	Steel braced legs	3.0	$0.33S_{DS}I_eW$			
	Steel cantilever legs	2.0	$0.50S_{DS}I_eW$			
	Concrete cantilever	2.0	$0.50S_{DS}I_eW$			

Table 4.C-1. Determination of Base Shear for Selected Structures. (Continued)

Equipment/Structure Types	Components	R	Base Shear Equation			Remarks
			$V = \frac{S_{DS}W}{R/I_e}$	$F_p = \frac{0.4a_p S_{DS} W_p (1+2z/h)}{R_p/I_p}$	$V = 0.35_{DS}I_eW$	
7.0 Boilers	Light steel framed wall	2.0	$0.50 S_{DS}I_eW$			Nonbuilding structure with structural system similar to building
Transverse	or					
	Steel braced frame where bracing carries gravity (OCBF)	See	Table 4.C-2			
Longitudinal	or					
	Steel OMRF	See	Table 4.C-2			
	w/ height limits					
	w/ permitted height increase					
	w/ unlimited height					

Equipment/Structure Types	Components	R	Base Shear Equation			Remarks
			$V = \frac{S_{DS}W}{R/I_e}$	$F_p = \frac{0.4a_p S_{DS}W_p(1+2z/h)}{R_p/I_p}$	$V=0.35_{DS}I_e W$	
8.0 Cooling tower	Wooden braced frame	3.5	$0.29S_{DS}I_e W$			Nonbuilding structure with structural system similar to building

Transverse

Longitudinal

Table 4.C-2. Determination of Base Shear for Selected Structures.
Note: Refer to Tables 12.2-1 and 15.4-1 of ASCE 7-16 for footnotes associated with the framing systems listed in the table. Values of C_d and Ω_o can be found in ASCE 7 Tables 12.2-1 and 15.4-1.

Nonbuilding Structure Type	Detailing Requirements	R	Base Shear Equation $V= (S_{DS}W) / (R/I_e)$	Structural System and Height Limits A&B	C	D	E	F
Structural steel systems not specifically detailed for seismic resistance excluding cantilever column systems	AISC 360 (2016)	3	$0.33S_{DS}I_eW$	NL	NL	NP	NP	NP
Building frame systems:								
Steel eccentrically braced frames (EBF)	AISC 341 (2016)	8	$0.125S_{DS}I_eW$	NL	NL	160	160	100
Special steel concentrically braced frames	AISC 341	6	$0.167S_{DS}I_eW$	NL	NL	160	160	100
Ordinary steel concentrically braced frames (OCBF)	AISC 341	3.25	$0.31S_{DS}I_eW$	NL	NL	35	35	NP
With permitted height increase	AISC 341	2.5	$0.40S_{DS}I_eW$	NL	NL	160	160	100
With unlimited height	AISC 360	1.5	$0.67S_{DS}I_eW$	NL	NL	NL	NL	NL
Moment-resisting frame systems:								
Special steel moment frame (SMRF)	AISC 341	8	$0.125S_{DS}I_eW$	NL	NL	NL	NL	NL

Special reinforced concrete moment frame (SMRF)	ACI 318 (2014), incl. Ch. 18	8	$0.125S_{DS}I_eW$	NL	NL	NL	NL	NL
Intermediate steel moment frame (IMRF)	AISC 341	4.5	$0.22S_{DS}I_eW$	NL	NL	35	NP	NP
With permitted height increase	AISC 341	2.5	$0.40S_{DS}I_eW$	NL	NL	160	160	100
With unlimited height	AISC 341	1.5	$0.67S_{DS}I_eW$	NL	NL	NL	NL	NL
Intermediate reinforced concrete moment frame (IMRF)	ACI 318, incl. Ch. 18	5	$0.20S_{DS}I_eW$	NL	NL	NP	NP	NP
With permitted height increase	ACI 318, incl. Ch. 18	3	$0.33S_{DS}I_eW$	NL	NL	50	50	50
With unlimited height	ACI 318, incl. Ch. 18	0.8	$1.25S_{DS}I_eW$	NL	NL	NL	NL	NL
Ordinary steel moment frame (OMRF)	AISC 341	3.5	$0.29S_{DS}I_eW$	NL	NL	NP	NP	NP
With permitted height increase	AISC 341	2.5	$0.40S_{DS}I_eW$	NL	NL	100	100	NP
With unlimited height	AISC 360	1	$1.0S_{DS}I_eW$	NL	NL	NL	NL	NL
Ordinary reinforced concrete moment frame (OMRF)	ACI 318, excl. Ch. 18	3	$0.33S_{DS}I_eW$	NL	NP	NP	NP	NP
With permitted height increase	ACI 318, excl. Ch. 18	0.8	$1.25S_{DS}I_eW$	NL	NL	50	50	50

APPENDIX 4.D STABILITY CHECK USING ENERGY BALANCE APPROACH (EXISTING FACILITIES ONLY)

4.D.1 INTRODUCTION

The factor of safety against overturning under static loading is defined as the ratio of the resisting forces to the overturning forces. However, this classical approach may be overly conservative when used for earthquake loadings because it does not recognize the dynamic character of the loading. The methodology described in this appendix is adapted from material presented by Bechtel (1980) and presents an approach based on energy balance. In this method, the factor of safety against overturning during earthquake loading is defined as the ratio of potential energy (Pe_o) required to cause overturning about one edge of the structure to the maximum kinetic energy (Ke_s) in the structure due to the earthquake. This factor of safety should be at least 1.5. Presented below is an overview of the methodology.

4.D.2 CALCULATION OF FACTOR OF SAFETY AGAINST OVERTURNING

The structure is considered unstable in the overturning mode when the amplitude of rocking motion causes the center of structural mass to reach a position over any edge of the base (Figure 4.D-1). The mechanism of rocking motion is that of an inverted pendulum with a very long natural period compared with that of the linear, elastic structural response. Hence, so far as overturning evaluation is concerned, the structure can be treated as a rigid body.

The factor of safety against overturning (FS) is given by

$$FS = \frac{PE}{KE}$$

where

PE = Potential energy to overturn the structure

= $mg\Delta h$

KE = Kinetic energy of earthquake input to the structure

= $\frac{1}{2}mV^2$

m = total mass of the structure

g = acceleration due to gravity

V = resultant total velocity of the structure

Δh = height to which the center of mass of the structure must be lifted to reach overturnin g position for a block as shown in Fig. 4D.1, Δh is given by:

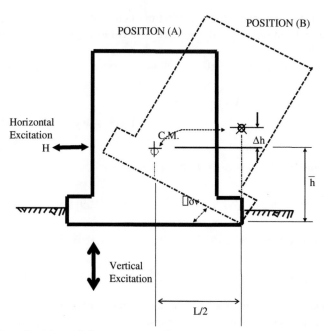

Figure 4.D-1. Position of the structure when overturning about one edge.

$$\Delta h = \left(h^2 + \left(\frac{L}{2} \right)^2 \right)^{1/2} - h$$

h = height of center of mass
$L/2$ = half width of shorter side of structure
    The factor of safety, FS, is then given by

$$FS = \frac{mg\,\Delta h}{1/2mV^2} = \frac{2g\,\Delta h}{V^2}$$

The velocity, V, is calculated as follows:

Total vertical velocity, $Vv = (Vvg^2 + Vvs^2)^{1/2}$

Total x-direction velocity, $Vx = (Vxg^2 + Vxs^2)^{1/2}$

Total y-direction velocity, $Vy = (Vyg^2 + Vys^2)^{1/2}$

where
    Vvg, Vxg, and Vyg are the peak ground velocities in the vertical, x, and y directions, and
    Vvs, Vxs, and Vys are the structure velocities in the vertical, x, and y directions.

Using the component factor method to obtain total structure response from separate lateral and vertical analyses, three different load combinations to consider are as follows:

Combination 1: 100% Vertical + 30% Horizontal x + 30% Horizontal y

$$V = \left[Vv^2 + (0.3Vx)^2 + (0.3Vy)^2 \right]^{1/2}$$
$$= \left[Vvg^2 + Vvs^2 + 0.09(Vxg^2 + Vxs^2 + Vyg^2 + Vys^2) \right]^{1/2}$$

Combination 2: 30% Vertical + 100% Horizontal x + 30% Horizontal y

$$V = \left[0.09(Vvg^2 + Vvs^2) + Vxg^2 + Vxs^2 + 0.09(Vyg^2 + Vys^2) \right]^{1/2}$$

Combination 3: 30% Vertical + 30% Horizontal x + 100% Horizontal y

$$V = \left[0.09(Vvg^2 + Vvs^2 + Vxg^2 + Vxs^2) + Vyg^2 + Vys^2 \right]^{1/2}$$

For cases where the specified x and y horizontal ground motions and structural responses are the same (i.e., conservative use of the peak of the velocity response spectra), Combinations 2 and 3 are identical and hence only two load combinations need to be considered.

Note that in the above computation of total vertical and horizontal velocities, the true relative velocity is more appropriate to use than the peak ground velocity. However, for long-period oscillators, this value approaches the ground velocity, whereas the commonly used pseudo-spectral velocity tends to zero. Therefore, to guard against the use of velocities that are below the ground velocity, the peak ground velocity has been added to the estimates of the peak structural velocity. No penalty is associated with this conservative approach because the factor of safety is ordinarily a very large number, despite the use of conservative peak velocities. The peak horizontal ground velocity can be obtained from any appropriate source. Lacking other information, a value of 4 ft/s for 1g maximum ground acceleration is recommended for competent soils (Newmark and Hall 1982). For the vertical ground motion, the peak ground velocity shall be 2/3 of this value. Structure velocities can be obtained knowing the structural periods and the appropriate acceleration response spectra in all three directions.

4.D.3 OTHER EFFECTS

The above analysis assumes rigid body rotation about one edge of the structure that is not free to work. Other effects such as soil flexibility, embedment, and buoyancy can significantly influence the computed factor of safety depending upon local conditions.

4.D.4 EFFECT OF SOIL FLEXIBILITY

Considering soil flexibility has two significant effects. First, it changes the axis about which "effective" rotation occurs. The axis of rotation moves inboard from the edge of the structure, as Figure 4.D-2 shows, toward the center, thereby making it easier to rotate the structure toward instability. Second, the soil can now absorb energy (because it has flexibility). The amount of energy absorbed can be significant, and this effect adds to the "resistance" to overturning. The instability equation now becomes

$$KE = PE + SE$$

where
SE = strain energy of the soil block

A factor of safety of 1.5 is now required on the angle causing overturning versus the angle at which the above equation holds true (i.e., $\theta ov/\theta \geq 1.5$). A check of the computed maximum soil pressure against the allowable bearing capacity should also be performed.

4.D.5 EFFECT OF EMBEDMENT

Embedment gives rise to additional resistance against overturning due to the side passive soil pressure considered as follows.

Let d be the depth of embedment and d' be the submerged depth where the ground water table is above the elevation of the base. The structure is assumed to rotate about the toe edge for the overturning evaluation. To simplify the analysis for practical purposes, only the passive soil pressure developed on the toe side is considered, and the wall friction and action of the soil on the opposite side of the structure are neglected. The passive pressure diagram is modified to be consistent with the assumption that the structure rotates about one edge. Free-draining soil conditions are also assumed. Figures 4.D.3(a) through (c) show the resultant idealized passive pressure diagram (P_z) for different elevations of the ground water table when it is above the base (i.e., $d' \geq 0$).

As was the case when considering soil flexibility at the base of the structure, the effect of embedment is to include some strain energy absorbed by the soil in the instability equation $KE = PE + SE$. Calculation of the strain energy can be performed through integration of the pressure and strain distribution over the volume of soil affected.

4.D.6 EFFECT OF BUOYANCY

When the ground water table is above the base ($d' > 0$), the buoyant force has the effect of increasing the structure's overturning potential. The buoyant force B acts

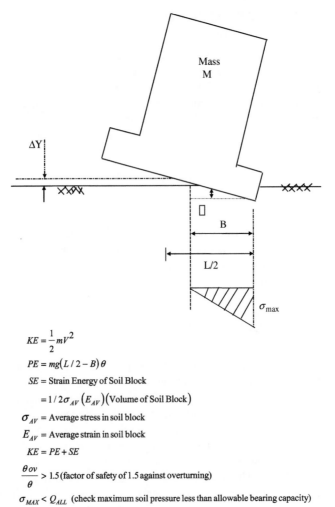

$$KE = \frac{1}{2}mV^2$$

$$PE = mg(L/2 - B)\theta$$

SE = Strain Energy of Soil Block

$$= 1/2\sigma_{AV}(E_{AV})(\text{Volume of Soil Block})$$

σ_{AV} = Average stress in soil block

E_{AV} = Average strain in soil block

$$KE = PE + SE$$

$\dfrac{\theta_{ov}}{\theta} > 1.5$ (factor of safety of 1.5 against overturning)

$\sigma_{MAX} < Q_{ALL}$ (check maximum soil pressure less than allowable bearing capacity)

Figure 4.D-2. Effect of soil flexibility.

at the centroid of the volume of water displaced by the submerged portion of the structure, and its magnitude varies during the overturning process. At any position before overturning takes place, the centroid of the displaced volume of water is located at height z above the elevation of the edge R and the corresponding buoyant force is B_z (Figure 4.D-4). The work done by the buoyant force is

$$W_B = \int_{z_a}^{z_b} B_z dz$$

where z_a and z_b are the heights of the centroid of buoyant force above the edge for the equilibrium and tipping positions, respectively.

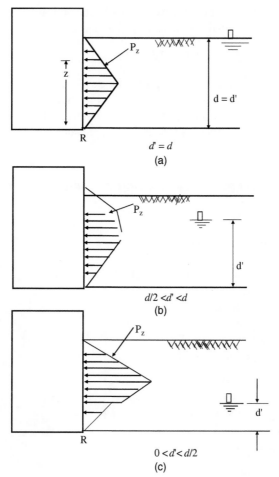

Figure 4.D-3. *Idealized passive soil pressure for overturning about edge* R.

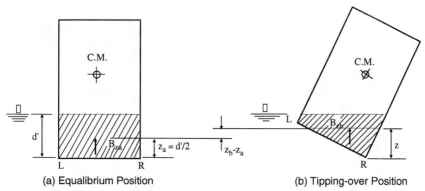

Figure 4.D-4. *Buoyancy effects for overturning evaluation.*

In the instability equation, the work done by the buoyant force is additive to the kinetic energy imposed by the earthquake, that is,

$$KE + W_B = PE + SE$$

In addition to buoyancy effects, if the block is sitting on saturated ground, suction is created when the block tries to lift off. Note that the positive effect of this suction in resisting overturning is difficult to quantify, but may be much larger than any effects of buoyancy, especially when the water is very shallow.

EXAMPLE 4D: STABILITY CHECK USING ENERGY BALANCE APPROACH

Given: Vertical Process Vessel
　　　9 ft diam. × 132 ft high (incl. skirt)
　　　Shell thickness: 5/8 in. thick
　　　Skirt: 9 ft diam. × 14 ft
　　　　high × 1-1/8 in. thick
　　　Empty weight: 343 kips
　　　Operating weight: 428 kips
　　　Structural period, $T = 1$ s

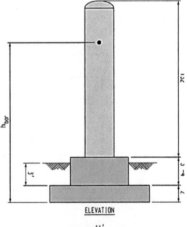

ELEVATION

　　　Soil and Foundation
　　　Concrete pedestal 11 ft
　　　　octagon × 3 ft-4 in. deep
　　　Concrete footing 25 ft
　　　　octagon × 2 ft deep

　　　Seismic Response Spectra
　　　PGA = 0.38 g
　　　$S_a = 0.45$ g at $T = 1.0$ s

(a) Determine seismic height (h_{bar}):
　　　Vessel:
　　　From column lateral analysis:
　　　Seismic center, $H_s = 89$ ft above
　　　　base plate
　　　Pedestal weight (W_p):

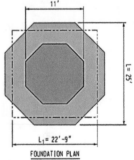

FOUNDATION PLAN

$$W_p = \left(\frac{11}{2}\right)^2 \tan(22.5)(8)(3.33)(0.15) = 50.07 \text{ kips}$$

Footing weight (W_f):

$$W_f = \left(\frac{25}{2}\right)^2 \tan(22.5)(8)(2)(0.15) = 155.33 \text{ kips}$$

Weight of soil above footing (W_s):

$$W_s = \tan(22.5)\left[\left(\frac{25}{2}\right)^2 - \left(\frac{11}{2}\right)^2\right](8)(3)(0.100) = 125.26 \text{ kips}$$

Total weight, $W_T = 428 + 50.07 + 155.33 + 125.26 = 758.66$ kips

Seismic height, h_{bar}

$$= \frac{428(89 + 5.33) + 50.07\left(\frac{3.33}{2} + 2\right) + 155.33(1) + 125.26\left(\frac{3}{2} + 2\right)}{758.66}$$

$$= 54.24 \text{ ft}$$

(b) Calculate potential energy (PE):

$$L_1 = L_2 = \left[\left(\frac{25}{2}\right)^2 \tan(22.5)(8)\right]^{0.5} = 22.75 \text{ ft}$$

$$\Delta h = \left[(h_{bar})^2 + \left(\frac{L}{2}\right)^2\right]^{0.5} - h_{bar} = \left[54.24^2 + \left(\frac{22.75}{2}\right)^2\right]^{0.5}$$

$$- 54.24 = 1.18 \text{ ft}$$

$PE = W_T \times \Delta h = 758.66\,(1.18) = 895.22$ kip-ft
Overturning rotation, $\theta_{ov} = 2\Delta h/L_1 = 2(1.18)/22.75 = 0.104$ radian

(c) Calculate kinetic energy (KE):
   $m = 758.66/32.2 = 23.56$ kip-s^2/ft
   From the seismic response spectra at the site, the vessel's horizontal acceleration at $T = 1.0$ s is $S_a = 0.45$ g.
   Vessel velocity, $V_{xs} = V_{ys} = S_a T/(2\pi) = 0.45(32.2)(1)/(2\pi) = 2.31$ ft/s
   The ground velocity is equal to 4 ft/s per 1.0g ground acceleration.
   From the seismic response, ZPA $= 0.38$ g
   Ground horizontal velocity, $V_{xg} = V_{yg} = 4(0.38) = 1.52$ ft/s
   Ground vertical velocity, $V_{vg} = 2/3\ V_{xg} = (2/3)1.52 = 1.01$ ft/s
   Calculate the resultant velocity response of the ground and vessel using the component factor method.

   From Combination 1:
   $V_1 = [V_{vg}^2 + V_{vs}^2 + 0.09(V_{xg}^2 + V_{xs}^2 + V_{yg}^2 + V_{ys}^2)]^{0.5}$
   $V_1 = [1.01^2 + 0 + 0.09(1.52^2 + 2.31^2 + 1.52^2 + 2.31^2)]^{0.5} = 1.55$ ft/s

   From Combination 2:
   $V_2 = [0.09(V_{vg}^2 + V_{vs}^2) + (V_{xg}^2 + V_{xs}^2) + 0.09(V_{yg}^2 + V_{ys}^2)]^{0.5}$
   $V_2 = [0.09(1.01^2 + 0) + (1.52^2 + 2.31^2) + 0.09(1.52^2 + 2.31^2)]^{0.5} = 2.90$ ft/s

From Combination 3:
$$V_3 = [0.09(V_{vg}^2 + V_{vs}^2) + 0.09(V_{xg}^2 + V_{xs}^2) + (V_{yg}^2 + V_{ys}^2)]^{0.5}$$
$$V_3 = [0.09(1.01^2 + 0) + 0.09(1.52^2 + 2.31^2) + (1.52^2 + 2.31^2)]^{0.5} = 2.90 \text{ ft/s}$$

Therefore, total velocity, $V = 2.90$ ft/s.
$$KE = 0.5mV^2 = 0.5(23.56)(2.90)^2 = 99.01 \text{ kip-ft}$$

(d) Compare PE and KE:
$$PE/KE = 895.22/99.01 = 9.04 > 1.5 \text{ OK!}$$

In steps (a) through (d) above, the strain energy owing to the flexibility of the underlying soil has not yet been considered. The overturning potential evaluation may be concluded here if the soil is infinitely rigid, such as rock. However, to illustrate a complete procedure to calculate the overturning potential for a vessel founded on flexible soil, this example is continued by assuming the soil beneath the foundation mat to be sand with the following parameters:

Soil parameter: $E_s = 1,000$ ksf; $Q_{ult} = 10$ ksf
                  $\mu = 0.3$ $Q_{all} = 5$ ksf

(e) Calculate the potential energy, which includes the soil deformation:

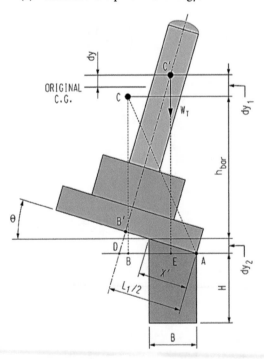

X' = Effective bearing length, ft
H = Effective depth of soil block under bearing pressure
L_1 = Shorter dimension of rectangular footing or equivalent square of octagon footing

L_2 = Longer dimension of rectangular footing or equivalent square of octagon footing

$B = X' \cos\theta$

dy = Change in elevation

$AD = AB'/\cos\theta = L_1/(2\cos\theta)$ $C'D = h_{bar} + (L_1/2)\tan\theta$

$B'D = [AD^2 - AB'^2]^{0.5}$ $C'E = C'D\cos\theta = dy1 + h_{bar}$

$\quad = [(L_1/2\cos\theta)^2 - (L_1/2)^2]^{0.5}$ $dy1 = [h_{bar} + (L_1/2)\tan\theta]\cos\theta - h_{bar}$

$\quad = L_1/2(1/\cos^2\theta - 1)^{0.5}$ $\quad = h_{bar}(\cos\theta - 1) + (L_1/2)\sin\theta$

$\quad = (L_1/2)\tan\theta$ $dy2 = B\theta$

$dy = dy1 - dy2 = h_{bar}(\cos\theta - 1) - (L_1/2)\sin\theta - \beta\theta$

$\quad = h_{bar}(1 - \theta^2/2 - 1) + (L_1/2)\theta - \beta\theta$

$\quad = [(L_1/2) - B]\theta - (h_{bar}/2)\theta^2$

$$\boxed{PE' = W_T \times dy = W_T \times [(L_1/2)\theta - B\theta - (h_{bar}/2)\theta^2]} \qquad (I)$$

Note: $\sin\theta = \theta + \theta^3/3 + \ldots\ldots\ldots$

$\quad\quad \cos\theta = 1 - \theta^2/2 + \theta^4/4 - \theta^6/6 + \ldots\ldots$

$\quad\quad$ Neglecting higher terms due to small angle

$\quad\quad\quad \sin\theta = \theta$

$\quad\quad\quad \cos\theta = 1 - \theta^2/2$

(f) Calculate strain energy (SE) from soil deformation:

$SE = 0.5(q_{av})(\varepsilon_{av})V$

where

q_{av} = Average soil bearing pressure

$\quad$ (ksf) = $(q_{max}/2) = (E_s)(\varepsilon)/2 = (E_s/2)(B\theta/H) = B\theta = E_s/2H$

ε_{av} = Average soil strain (in./in.) = $q_{av}/E_s = B\theta/(2H)$

$\quad V$ = Volume of soil under pressure (ft^3) = BL_2H

$SE = 0.5(B\theta E_s/2H)(B\theta/2H)(BL_2H) \;\text{------}\!>\!>\! SE = B^3\theta^2 E_s \times L_2/8H$

The relationship between the soil strain (compressibility) and the rotation angle θ of the mat foundation can be determined as follows:

$X' = 3(L_1/2 - e)$ when $e > L_1/6$

$X' = L_1$ when $e < L_1/6$

$e = M_{OT}/W_T$ M_{OT} = Equivalent static moment

Note: It is assumed that $e > L_1/6$ for the extreme overturning case, otherwise this problem can be solved by the conventional static procedure.

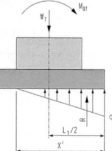

$R = 0.5\,q_{max}X'\,L_2$

$W_T - m(Sa_{vs} - Sa_{vg}) = R$ $Sa_{vs} = 0$ for (vertically) rigid vessel

$R = W_T + m\,Sa_{vg}$ $Sa_{vg} = (30\%)(2/3\ ZPA)$

$\quad\quad\quad\quad = 0.3(2/3)(0.38g) = 0.08\ g$

$R = 758.66(1 + 0.08) = 819.35$ kips

$R = 0.5q_{max}X' L_2$
$\quad = (0.5)(B\theta E_s/H)(B/\cos\theta)L_2$
$\quad = B^2\theta E_s L_2/2 H$

Rearranging this equation and solving for angle θ yields

$\theta = 2\ RH/B^2E_s \times L_2 = 2(819.35)H/[(B^2)(1,000)(22.5)] = \boxed{0.072H/B^2}$

From *Foundation Analysis and Design*, 5th ed., Sections 5–7 by Joseph E. Bowels (1996):

$\tan\theta = (1 - \mu^2)/E_s(M_{OT}/B^2L_2)I_\theta$

$\tan\theta = \theta$ for small angle
$2\ RH/B^2E_s L_2 = (1 - \mu^2)/E_s(M_{OT}/B^2L_2)I_\theta$
$2\ RH = (1 - \mu^2)M_{OT}I_\theta$
$H = [(1 - \mu^2)M_{OT}I_\theta]/2R$

Substituting $MOT = R\ (L_1/2 - X'/3)$ into the previous equation, we can solve for H as follows:

$H = (1 - \mu^2)(L_1/4 - X'/6) \times I_\theta$

Assuming the soil to be a dense sand and considering the octagon foundation to be an equivalent square footing, obtain the values of $I\theta$ from *Foundation Analysis and Design*, 5th ed., Table 5-5, by Joseph E. Bowels (1996).

$\mu = 0.3$ and $I_\theta = 4.2$

Therefore,

$$H = (1 - 0.3^2)\left(\frac{L_1}{4} - \frac{X'}{6}\right)$$
$$H = 3.822\left(\frac{L_1}{4} - \frac{X'}{6}\right)$$

(4.2)

Substituting $\theta = 2\ RH/B^2E_sL_2$ in the strain energy equation (SE),
$SE = B^3\theta^2E_sL_2/8H$
$\quad = B^3(0.072\ H/B^2)^2E_sL_2/8H$
$\quad = (0.072)^2H\ (1,000)\ 22.75/8B = \boxed{14.74(H/B)}$

(g) Balance the total kinetic, potential, and strain energies in the system:
$KE = PE' + SE$
$KE = 99.01$ kip-ft (calculated earlier)
$PE' = W_T[(L_1/2)\theta - B\theta - (h_{bar}/2)\theta^2]$
$\quad = 758.66\ [(22.75/2)(0.072\ H/B^2) - B\ (0.072\ H/B^2)$
$\quad\quad - (54.24/2)(0.072\ H/B^2)^2]$
$\quad = 621.34\ (H/B^2) - 54.62\ (H/B) - 106.66\ (H^2/B^4)$

$SE = 14.74 \ (H/B)$

$KE = PE' + SE = 621.34 \ (H/B^2) - 54.62 \ (H/B) - 106.66 \ (H^2/B^4)$
$\qquad + 14.74 \ (H/B)$

$621.34 \ (H/B^2) - 39.88 \ (H/B) - 106.66 \ (H^2/B^4) = 99.006 \ \text{kip-ft} \qquad (A)$

$H = 3.822 \ (L_1/4 - X'/6)$ (calculated earlier)

For a small angle θ, $B = X' = 3(L_1/2 - e)$
$\qquad\qquad\qquad\qquad\quad = 3(L_1/2 - M_{OT}/R)$
$\qquad\qquad\qquad\qquad\quad = 3(22.75/2 - M_{OT}/819.35)$
$\qquad\qquad\qquad\qquad\quad = 34.13 - M_{OT}/273.12 \qquad (B)$

$H = 3.822 \ (L_1/4 - L_1/4 + M_{OT}/2R)$
$\quad = 3.822 \ M_{OT}/[(2)(819.35)]$
$\quad = 2.332 \times 10^{-03} M_{OT} \qquad (C)$

For a range of $L_1/6 < e < L_1/2$, select various M_{OT} values and solve for H, B, and KE by trial and error.

Solution: $M_{OT} = 7355.75$ kip-ft
$B = X' = 3[(22.75/2) - (7355.75/819.35)] = 7.192$
$H = [3.822(7355.75)]/[2(819.35)] = 17.156$
$e = M_{OT}/R = 7355.75/819.35 = 8.978 > L_1/6 = 3.78$ ft and
$\qquad < L_1/2 = 11.36$ ft   OK!
$KE = 621.34(17.156/7.192^2) - 39.88(17.156/7.192) - 106.66(17.156^2/7.192^4)$
$\qquad = 99.22 \ \text{kip-ft} \sim 99.01 \ \text{kip-ft}$ OK!
$\theta = 0.072 \ (H/B^2) = 0.072(17.156)/7.192^2 = 0.024$ radians
$q_{max} = 2R/[(X')(L_2)] = 2(819.35)/[7.192(22.75)]$
$\qquad = 10.02 \ \text{ksf} \sim Q_{ult} = 10 \ \text{ksf}$ OK!

Summary:
$PE/KE = 9.042 > 1.5 \qquad$ OK!
$q_{max} = 10.02 \ \text{ksf} \sim Q_{ult} = 10 \ \text{ksf} \qquad$ OK!
$\theta_{ov}/\theta = 0.104/0.024 = 4.33 > 1.5 \qquad$ OK!
$L_1/2 < e < L_1/6 \qquad$ OK!

APPENDIX 4.E METHODOLOGY FOR DETERMINATION OF SLIDING DISPLACEMENTS (EXISTING FACILITIES ONLY)

E.1 INTRODUCTION

The occurrence of sliding of the structure as a rigid body due to earthquake loads does not necessarily lead to a failure mode as long as sliding is within tolerable limits. A measure of the factor of safety then is the ratio of the calculated to allowable displacements. The methodology described in this appendix is adapted from material presented by Bechtel (1980) that follows a procedure originally developed by Newmark, in which an acceleration, ng, that would just cause a rigid block to slide is defined.

E.2 METHODOLOGY

Assuming a single acceleration pulse of magnitude $a_h g$ and of duration t_1 is applied to the rigid block (Figure 4.E-1b), a velocity plot, as in Figure 4.E-1c, can be constructed. Given the maximum horizontal ground velocity, V_{Hmax}, time t_1 is then estimated from

$$t_1 = \frac{V_{H\,max}}{a_h g}$$

The velocity due to the resisting acceleration is given by ngt. At time t_2, the two velocities are equal, and the rigid block comes to rest relative to the ground. From Figure 4.E-1c, time t_2 is given by

$$t_2 = \frac{V_{H\,max}}{ng}$$

The maximum displacement of the rigid block relative to the ground is obtained by integrating the velocity curves up to time t_2 and subtracting the results. This is equivalent to computing the area of the shaded triangle. The maximum is then given by

$$\Delta_s = \frac{V_{H\,max}^2}{2ng}\left(1 - \frac{n}{a_h}\right)$$

where

Δ_s = Maximum sliding displacement,

V_{Hmax} = Maximum total velocity of the structure in the horizontal direction,

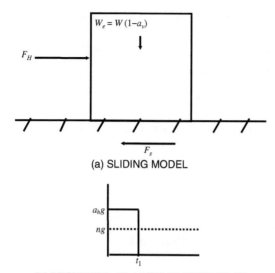

(a) SLIDING MODEL

(b) RECTANGULAR ACCELERATOR PULSE

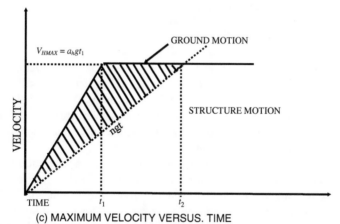

(c) MAXIMUM VELOCITY VERSUS. TIME

Figure 4.E-1. Sliding because of earthquake.

g = Acceleration owing to gravity,

a_h = Maximum total horizontal acceleration in g,

n = Fraction of gravitational acceleration that would cause the structure to slide,

= $\mu\,(1 - a_v)$

μ = Coefficient of friction between structure and ground, and

a_v = Maximum vertical acceleration.

Using the component factor method, V_{Hmax}, a_h, and n are calculated as follows:

Combination 1: 100% Vertical + 30% Horizontal x + 30% Horizontal y

$$V_{H\,max} = [0.09(Vxg^2 + Vxs^2 + Vyg^2 + Vys^2)]^{1/2}$$
$$a_h = [0.09(a_x^2 + a_y^2)]^{1/2}$$
$$n = \mu(1 - a_v)$$

Combination 2: 30% Vertical + 100% Horizontal x + 30% Horizontal y

$$V_{H\,max} = [(Vxg^2 + Vxs^2 + 0.09(Vyg^2 + Vys^2)]^{1/2}$$
$$a_h = [a_x^2 + 0.09a_y^2]^{1/2}$$
$$n = \mu(1 - 0.3a_v)$$

Combination 3: 30% Vertical + 30% Horizontal x + 100% Horizontal y

$$V_{H\,max} = [Vyg^2 + Vys^2 + 0.09(Vxg^2 + Vxs^2)]^{1/2}$$
$$a_h = [a_y^2 + 0.09a_x^2]^{1/2}$$
$$n = \mu(1 - 0.3a_v)$$

In the above equations Vxg, Vyg, and Vvg are the peak ground velocities in the x, y, and vertical directions; Vxs, Vys, and Vvs are the structure velocities in the x, y, and vertical directions; and a_x, a_y, and a_v are the peak amplified structural accelerations in the x, y, and vertical directions. Vertical ground response parameters can be taken as 2/3 of horizontal values, and values of peak ground velocity can be estimated from empirical equations.

Note from the equation for Δ_s that when the applied acceleration coefficient a_h is less than or equal to the resisting acceleration coefficient n, Δ_s equals zero. This result is compatible with the definition of the coefficient n. Relative displacement would occur only when a_h is greater than n. Setting n/a_h in the equation equal to zero will provide an upper-bound estimate of Δ_s; however, this simplification leads to finite displacements for all cases.

Note that to calculate total displacement, the sliding displacement needs to be added to the in-structure deformation, assuming the structure does not move at its base.

APPENDIX 4.F

GUIDANCE FOR CALIFORNIA ACCIDENTAL RELEASE PREVENTION (CalARP) PROGRAM

SEISMIC ASSESSMENTS

Committee Note: This appendix contains a document prepared by others for use in evaluating existing facilities. It is reprinted in its entirety because it is currently being used by regulators throughout California for facilities containing Acutely Hazardous Materials, but it is not published and is not easily obtainable by practicing engineers.

GUIDANCE
FOR
CALIFORNIA
ACCIDENTAL RELEASE PREVENTION (CalARP) PROGRAM
SEISMIC ASSESSMENTS

Prepared for the
ADMINISTERING AGENCY (AA) SUBCOMMITTEE
REGION I LOCAL EMERGENCY PLANNING COMMITTEE
(LEPC)

Prepared by the
CalARP PROGRAM SEISMIC GUIDANCE COMMITTEE
December 2013

Approved by Region I LEPC on March 12, 2014

CalARP PROGRAM SEISMIC GUIDANCE COMMITTEE

CONSULTANTS

Robert Bachman	R. E. Bachman, Consulting Structural Engineer–Committee Chair
Brian Hopper	Hopper Engineering Associates
Gayle Johnson	Simpson Gumpertz & Heger
Guzhao Li	MMI Engineering, Inc.
Krishna Nand	Environmental Management Professionals
Kenneth Saunders	Argos Engineers, Inc. – Secretary
Paul Summers	MMI Engineering, Inc.
Darius Vitkus	Gekko Engineering, Inc.
Derrick Watkins	Tobolski Watkins Engineering, Inc.
Curtis Yokoyama	Fluor Enterprises, Inc.
Farzin Zareian	University of California - Irvine

INDUSTRY REPRESENTATIVES

Paul Beswick	Metropolitan Water District of Southern California - LEPC AA Member
Vincent Borov	Chevron ETC. – San Ramon
Winston Chai	Metropolitan Water District of Southern California
Ronald Orantes	Los Angeles Department of Water and Power
Andrew Mitchell	Chevron/Enova
Rocco Serrato	Chevron - El Segundo

AGENCY REPRESENTATIVES

Richard Clark	Los Angeles County Fire Department
Mashid Harrell	Los Angeles County Fire Department
Richard Kallman	Santa Fe Springs Fire Department
Jerrick Torres	City of Vernon - Health & Environmental Control Department

TABLE OF CONTENTS

1.0 INTRODUCTION

The objective of a California Accidental Release Prevention (CalARP) Program seismic assessment is to provide reasonable assurance that a release of Regulated Substances (RS) as listed in California Code of Regulations (CCR) Title 19 Division 2 Chapter 4.5 (Reference 1) having offsite consequences (caused by a loss of containment or pressure boundary integrity) would not occur as a result of an earthquake. Since 1998, the seismic assessment study has been part of the mandated State's CalARP program. The purpose of this document is to provide guidance regarding criteria to be used in such assessments. This guidance document is an update of the CalARP seismic document published in September of 2009 (Reference 2). The guidance provided is applicable to structural systems and components whose failure could result in the release of sufficient quantities of RS to be of concern.

The guidance given in this document provides for a deterministic evaluation of structural systems and components. This deterministic evaluation should be performed considering an earthquake which has a low probability of occurrence (code Design Earthquake level as defined in ASCE/SEI 7-10 (Reference 4)). The seismic capacity of structures and components to withstand this level of earthquake should be calculated using realistic criteria and assumptions.

An acceptable alternate approach is to perform a probabilistic risk assessment which provides estimates and insights on the relative risks and vulnerabilities of different systems and components from the impact of an earthquake. These risks should be compatible with accepted practices for similar civil and industrial facilities. When a probabilistic risk assessment approach is planned, the owner/operator should consult with the local Authority(ies) Having Jurisdiction (AHJ) to describe why this approach is being planned and explain differences between this approach and the deterministic method.

The AHJ is usually State Unified Program Agencies (UPAs), also referred to as Administering Agencies (AA), that enforce CalARP program regulations; and may also include city or county Building & Safety Departments that approve plans and issue permits for renovation and/or construction/installation of structural systems/components.

The CalARP regulation states in Section 2760.2 (b): "The owner or operator shall work closely with AAs in deciding which PHA [Process Hazard Analysis] methodology is best suited to determine the hazards of the process being analyzed." Thus, prior to the beginning of any seismic assessment, the owner/operator needs to consult closely with the AHJ to obtain mutual understanding and agreement on the scope of the assessment, the general approach proposed by the Responsible Engineer (see Section 1.5) and the schedule for the assessment.

1.1 Limitations – Conformance to this document does not guarantee or assure that a RS release will not occur in the event of strong earthquake ground motions. Rather, the guidance provided is intended to reduce the likelihood of release of RS.

1.2 Evaluation Scope – The owner/operator, in consultation with the AHJ and Responsible Engineer (see Section 1.5), should always identify the systems to be evaluated in accordance with this guidance. The systems are expected to fall into three categories. These are:

1) Covered processes as defined by CalARP Program regulations.

2) Adjacent facilities whose structural failure or excessive displacement could result in the significant release of RS.

3) Onsite utility systems and emergency systems which would be required to operate following an earthquake for emergency reaction or to maintain the facility in a safe condition, (e.g., emergency power, leak detectors, pressure relief valves, battery racks, release treatment systems including scrubbers or water diffusers, firewater pumps and their fuel tanks, cooling water, room ventilation, etc.).

1.3 Performance Criteria – In order to achieve the overall objective of preventing releases of RS, individual equipment items, structures, and systems (e.g., power, water, etc.) may need to achieve varied performance criteria. These criteria may include one or more of the following:

1) Maintain structural integrity

2) Maintain position

3) Maintain containment of material

4) Function immediately following an earthquake

Note that an owner/operator may choose to set more stringent performance requirements dealing with continued function of the facilities both during and after an earthquake. These are individual business decisions and are not required for compliance with the CalARP Program.

From January 1, 2008 to December 31, 2013, all new facilities in California should have been designed in accordance with the 2007 California Building Code (CBC) which references the 2006 International Building Code (IBC) seismic requirements or the 2010 CBC which references the 2009 IBC. Both the 2006 IBC and the 2009 IBC in turn reference American Society of Civil Engineers (ASCE) Standard ASCE/SEI 7-05 for their seismic load provisions. Starting on January 1, 2014, all new facilities in California are to be designed in accordance with the 2013 CBC (Reference 22) which references the 2012 IBC (Reference 19) seismic requirements. The 2012 IBC in turn references American Society of Civil Engineers (ASCE) Standard ASCE/SEI 7-10 (Reference 4) for its seismic load provisions. It is the consensus of this Committee that RS systems and components designed and properly constructed in accordance with the 1997 UBC (Reference 3) or ASCE/SEI 7-05 (or later) provisions provide reasonable assurance of withstanding design/ evaluation basis earthquake effects without either structural failure or a release of RS having offsite consequences. It is also the consensus of this Committee that RS systems and components that were designed and constructed in accordance with the 1988, 1991 or 1994 UBC also provide reasonable assurance of withstanding design/

evaluation basis earthquake effects without either structural failure or a release of RS (caused by a loss of containment or pressure boundary integrity) provided that the facility in which the systems and components are contained is not located in the near field of an active earthquake fault or on a soft soil site. It should be noted that design earthquake terminology changed between the UBC and ASCE/SEI 7-05 and between ASCE/SEI 7-05 and ASCE/SEI 7-10 The design earthquake ground motion level in the UBC is called the "design basis earthquake" while in ASCE/SEI 7-05 and ASCE/SEI 7-10, it is called "design earthquake" (DE). Also, the maximum earthquake ground motion considered in ASCE/SEI 7-05 was called the "Maximum Considered Earthquake" (MCE) while the maximum earthquake ground motion considered in ASCE/SEI 7-10 is called the "Risk-Targeted Maximum Considered Earthquake" (MCE$_R$).

State and national policies have generally established performance objectives for new facilities that are more restrictive than those for existing facilities. This guidance document recognizes this to be appropriate. However, it should be recognized that any regular inspection and repair of systems containing RS should make them significantly safer than similar systems for which these steps are not taken.

1.4 Extent of Seismic Evaluations Required – All equipment and components identified in Section 1.2 are subject to the seismic assessment guidelines of this document. However, the extent of these evaluations may be limited or expanded depending on the situation. Each owner/operator will have different conditions at their facility and should consult with the AHJ to determine which of the following subsections apply to their facility.

1.4.1 Existing Facilities Which Have Not Had Previous CalARP Seismic Assessments

1) Constructed to 1985 UBC and Earlier
   There is considerable uncertainty about the capacity of nonbuilding structures and nonstructural components designed and constructed prior to the 1988 UBC. This is because there were no specific seismic code requirements for nonbuilding structures and nonstructural components in heavy industrial applications and they were rarely reviewed and inspected by building departments. Starting with the 1988 UBC, seismic code requirements were provided and designs were much more consistent. Therefore, pre-1988 UBC nonbuilding structure and nonstructural component designs should always be considered suspect and subject to CalARP type evaluations if they are in the evaluation scope (Section 1.2).

2) Constructed to 1988 UBC and Later
   Existing facilities which are subject to the CalARP requirements and which were permitted for construction in California in accordance with the 1988 or later version of the UBC may generally be deemed to meet the intent of the requirements of Section 4 of this Guidance, provided the following conditions are met and documented:

a. The near field requirements of either ASCE/SEI 7-05, ASCE/SEI 7-10 or the 1997 UBC, either using the near field maps or a site-specific spectrum, are satisfied or the facility is not located in the near field zone (i.e., where per ASCE/SEI 7-05 S_S is not greater than 1.5 and S_1 is not greater than 0.6 or per the 1997 UBC the facility is not within 15 km of an active fault).

b. The soft soil site conditions of ASCE/SEI 7-05 or the 1997 UBC were considered in the design of the facility or the facility is not located on a soft soil site.

c. A walkdown in accordance with Section 3 reveals adequate lateral force resisting systems.

The recommended contents of the initial report are given in Section 9.1.

1.4.2 New Facilities Submitted for Permit After December 2013 That Are Subject to CalARP Program Requirements – Design and construction of new facilities containing RS must satisfy the seismic provisions of the 2007 California Building Code (ASCE/SEI 7-05). In general, such facilities are deemed to satisfy the analytical evaluation requirements of the guidance document. However, a walkdown should always be performed in accordance with Section 3 after construction has been completed. The recommended contents of the initial report are given in Section 9.1.

1.4.3 Facility Revalidation With a Previous CalARP Seismic Assessment – The CalARP program requires that facilities, which are subject to the CalARP requirements, have their process hazard analysis updated and revalidated at least every five years. The extent of a seismic assessment revalidation depends on many factors that need to be coordinated and agreed to by the AHJ. If deemed appropriate by the Responsible Engineer (see Section 1.5), any portion of the previous assessment may be used for the current assessment. However, any revalidation should include the performance of a walkdown in accordance with Section 3 of this document. As part of the revalidation process the equipment population being assessed should be discussed with the process engineer responsible for defining the scope of the assessment. It is possible that process conditions have changed since the initial screening of equipment having offsite consequences was first performed.

The recommended contents of the revalidation report are given in Section 9.2.

1.4.4 Occurrence of Conditions That Would Trigger an Assessment Within the Revalidation Period – It is recommended that owners/operators assessing the validity of past evaluations consider conditions that may make a partial or entirely new assessment necessary. Examples of such conditions include: Major increases in the estimated ground motions (new significant active fault discovered near the facility)

1) Major increases in the estimated ground motions (e.g., new significant active fault discovered near the facility).

2) System modifications that would significantly affect the seismic behavior of the equipment or system, such as changing or addition of equipment or processes.

3) The occurrence of an earthquake that has caused significant damage in the local vicinity of the facility since the latest assessment.

4) The occurrence of other events (e.g., fire or explosion) that have caused structural damage.

5) Significant deterioration (e.g., corrosion) in equipment, piping, structural members, foundations or anchorages.

1.5 Responsible Engineer – The Responsible Engineer has responsibility for conducting and/or overseeing the evaluations and walkdowns required by this document for a given facility. All applicable engineering work associated with seismic evaluations should be performed or supervised by California Registered Professionals in accordance with the Business and Professions Code, Chapter 7, §§6700-6799 and CCR, Title 16, Division 5, §§400-476. It is strongly recommended that the Responsible Engineer be registered in California as a Civil, Structural or Mechanical Engineer with experience in seismic design and/or evaluations of facilities within the scope of this document.

2.0 DETERMINATION OF SEISMIC HAZARDS

When a seismic hazard assessment is performed, it should address and, where appropriate, quantify the following site-specific seismic hazards:

1) Ground shaking, including local site amplification effects
2) Fault rupture
3) Liquefaction and lateral spreading
4) Seismic settlement
5) Landslides
6) Tsunamis and seiches

Each of these site-specific seismic hazards is discussed in the following sections. Attachment A presents guidance for geotechnical reports that may be necessary to perform these evaluations.

2.1 Ground Shaking – It is the consensus of the Seismic Guidance Committee that the same ground motion hazard used in the design of new facilities be used as the basis for evaluating existing facilities. (i.e., the "Design Earthquake Response Spectrum" as per Section 11.4.5 of ASCE/SEI 7-10). The procedures of ASCE/SEI 7-10 should be used consistently for determination of these ground motions, including Chapter 21 of ASCE/SEI 7-10 for site-specific assessments. Values to be used in these evaluations may be obtained online from the United States Geological Survey (USGS) website at http://earthquake.usgs.gov//haz/designmaps/. Latitude and longitude of the facility should always be used, along with the appropriate soil classification.

2.2 Fault Rupture – Fault rupture zones which pass near or under the site should be identified. A fault is a fracture in the earth's crust along which the separated sections have moved or displaced in relation to each other. The displacement can be in either a horizontal or vertical direction. A ground rupture involving more than a few inches of movement can cause major damage to structures sited on the fault or pipelines that cross the fault. Fault displacements produce forces so great that the best method of limiting damage to structures is to avoid building in areas close to ground traces of active faults.

Under the Alquist-Priolo Special Studies Zones Act of 1972, the State Geologist is required to delineate "Earthquake Fault Zones" along known active faults in California. Fault maps are described and can be found online at the California Geological Survey (CGS) website at http://www.conserv.ca.gov/cgs/rghm/ap/index.htm in Special Publication 42 and the associated fault maps.

2.3 Liquefaction and Lateral Spreading – Liquefaction is the transformation of soil from solid to a liquid state caused by an increase in pore water pressure and a reduction of effective stress within the soil mass. The potential for liquefaction is greatest when loose saturated cohesionless (sandy) soils or silty soils of low plasticity are subjected to a long duration of seismically induced strong ground shaking.

The assessment of hazards associated with potential liquefaction of soil deposits should consider two basic types of hazards:

1) One type of hazard associated with liquefaction is translational site instability more commonly referred to as lateral spreading. Lateral spreading occurs on gently sloping ground with free-face (stream banks, and shorelines), when seams of liquefiable material are continuous over large lateral areas and serve as significant planes of weakness for translational movements.

2) Localized liquefaction hazards may include large liquefaction-induced settlements/differential settlements and foundation bearing failures.

The 2013 CBC, 2012 IBC and ASCE/SEI 7-10 require the liquefaction hazards be evaluated for the Maximum Considered Earthquake (MCE) geo-mean earthquake ground motions. Previous editions of the IBC and ASCE/SEI 7 required the liquefaction hazards be evaluated for the Design Earthquake (DE) hazard level.

It should be noted that although the new codes have changed their requirements regarding the hazard level, those changes are also associated with different performance expectations for the design of new structures (i.e., non-collapse) in the MCE. It is the consensus of this committee that changing the hazard levels for CalARP assessments of existing facilities to be consistent with philosophical changes in new design codes would add a level of complexity that is not justified and inconsistent with the approach used throughout this document. As such, this document continues to use the DE ground shaking levels to evaluate liquefaction hazards for existing facilities

The CGS has established evaluation guidelines in Special Publication 117 (SP117) (Reference 5). Preliminary screening investigations for liquefaction hazards should include the following:

1) Check the site against the liquefaction potential zone identified on the CGS Seismic Hazard Zones Maps where available.

2) Check for susceptible soil types. Most susceptible soil types include sandy soils and silty soils of low plasticity. Also susceptible are cohesive soils with low clay content (less than 15% finer than 0.005 mm), low liquid limit (less than 35%), and high moisture content (greater than 0.9 times the liquid limit). The latter may be designated as "quick" or "sensitive" clays.

3) Check for groundwater table. Liquefaction can only occur in susceptible soils below the groundwater table. Liquefaction hazards should be evaluated only if the highest possible groundwater table is shallower than 50 feet from the ground surface.

4) Check for in-situ soil densities to determine if they are sufficiently low to liquefy. Direct in-situ relative density measurements, such as the ASTM D 1586 (Standard Penetration Test) or ASTM D 3441 (Cone Penetration Test) or geophysical measurements of shear-wave velocities can provide useful

information for screening evaluation. This information will usually need to be evaluated by a geotechnical engineer.

The issue of liquefaction may be discounted if the geotechnical report or responsible engineer, using one or more of the above screening approaches, concludes that the likelihood of liquefaction is low.

A site-specific investigation and liquefaction evaluation may be omitted if a screening investigation can clearly demonstrate the absence of liquefaction hazards at site. Where the screening investigation indicates a site may be susceptible to liquefaction hazard, a more extensive site-specific investigation and liquefaction evaluation should be performed by a geotechnical engineer

2.4 Seismic Settlement – In addition to the effects of liquefaction, foundation settlement may occur due to soil compaction in strong ground shaking. A geotechnical engineer can determine the potential for this settlement.

2.5 Landslides – Facilities that are in close proximity to natural hillside terrain or man-made slopes (cut or fill slopes) are potentially susceptible to earthquake-induced landslide hazards. SP117 (Reference 5) presents guidelines for evaluation and mitigation of earthquake-induced landslide hazards. Information can also typically be obtained from the Seismic Safety Element of the General Plan. Preliminary screening investigation for such hazards should include the following:

1) As part of the site reconnaissance, the engineer should observe whether there are any existing slopes (natural or man-made) in the immediate vicinity of the facility.

2) If there are no slopes of significant extent within a reasonably adequate distance from the facility, then the potential for landslide may be dismissed as a likely seismic hazard. Engineering judgment may be used to assess what constitutes an "adequate distance." For example, generally level alluvial valleys can be reasonably excluded from the potential for seismically induced landslide.

3) If the facility is in close proximity to existing slopes which could pose a significant hazard, a certified engineering geologist or a registered geotechnical engineer should perform the following screening investigation steps.

   a. Check the site against the Seismic Slope Stability Hazard maps where available prepared by the CGS. Also check other similar maps from the USGS, Dibblee Geological Foundation (DGF), and Seismic Safety Elements of local cities and counties.

   b. Check the site against available published and unpublished geologic and landslide inventory maps.

   c. Review stereoscopic pairs of aerial photographs for distinctive landforms associated with landslides (steep slopes, scarps, troughs, disrupted drainages, etc.).

2.6 Tsunamis and Seiches

2.6.1 Background - Tsunamis, or tidal waves, are generated by distant earthquakes and undersea fault movement. Traveling through the deep ocean, a tsunami is a broad and shallow, but fast moving, wave that poses little danger to most vessels. When it reaches the coastline however, the waveform pushes upward from the ocean bottom to make a swell of water that breaks and washes inland with great force.

A seiche occurs when resonant wave oscillations form in an enclosed or semi-enclosed body of water such as a lake or bay. Seiches may be triggered by moderate or larger local submarine earthquakes and sometimes by large distant earthquakes. A tsunami or seiche may result in flooding of low-lying coastal areas. The greatest hazard results from the inflow and outflow of water, where strong currents and forces can erode foundations and sweep away structures and equipment. The rupture of storage tanks from debris impact and foundation erosion can result in fires and explosions.

In California, the Seismic Safety Elements of General Plans typically provide an estimate of the potential for tsunami and seiche inundation. Estimates of maximum tsunami run-up can be made using historical information or theoretical modeling.

Current methodologies for tsunami design are in development and are planned to be incorporated in the next edition of ASCE/SEI 7. These procedures will include methods to determine local tsunami risk, appropriate site design parameters, and procedures for analysis for tsunami loads and effects. Another useful resource for evaluating structures for tsunami loads is FEMA P646 (Reference 20). Although specifically intended for design of tsunami evacuation structures, the document presents analysis procedures and methods for determining tsunami loads which may be applicable to the structural systems and components within the scope of CalARP seismic assessments.

2.6.2 Administrative Mitigation Measures - Due in part to a lack of specific tsunami likelihood and/or probability of occurrence data, administrative mitigation measures are valuable. These include:

1) Early Warning System

2) Evacuation Planning

3) Hazardous Materials Area Plans and Regional Plans

4) Emergency Plant Shutdown Procedures

5) Coordination Emergency Drills

These measures would also be more achievable and timely than attempts to strengthen plant tankage and equipment from the effects of a large tsunami event.

2.6.3 Ongoing Developments for Mitigating Tsunami Hazards in the United States - The National Oceanographic and Atmospheric Administration (NOAA) is currently in the process of developing an early tsunami warning system for distant tsunami sources for the west coast of the United States. When the system becomes available, facilities which are vulnerable to a tsunami should be tied into the system and they should develop emergency plans in the event there is a tsunami warning issued that would affect their area.

"*Tsunami Risk Reduction for the United States: A Framework for Action*" (Reference 23), the joint report by the sub-committee on Disaster Reduction and the US Group on Earth Observations, called for development of a standardized and coordinated tsunami hazard and risk assessment for all coastal regions of the United States and its territories. In response to this report, at the request of the National Tsunami Hazard Mitigation Program (NTHMP), NOAA's National Geophysical Data Center (NGDC) and the United States Geological Survey (USGS) collaborated to conduct the first tsunami hazard assessment of the United States and its territories with the following conclusion: both the frequency and the amplitudes of tsunami run-ups support a qualitative "high" hazard assessment for Washington, Oregon, California, Puerto Rico, and the Virgin Islands. The "high" value for Oregon, Washington, and northern California reflects the low frequency but the potential for very high run-ups from magnitude 9 earthquakes on the Cascade subduction zone. Updates will be required as additional knowledge is obtained of possible tsunami sources in offshore southern California.

As the result of this review, the *National Tsunami Research Plan* (Reference 24) has been developed. The Plan has identified the following high priority research areas for improving the knowledge essential to tsunami risk reduction:

1) Enhance and sustain tsunami education

2) Improve tsunami warnings

3) Understand the impacts of tsunami at the coasts

4) Develop effective mitigation and recovery tools

5) Improve characterization of tsunami sources

6) Develop a tsunami data acquisition, archival, and retrieval system.

3.0 WALKDOWN CONSIDERATIONS

A critical feature of the evaluation methodology is the onsite review of the existing facility by a qualified engineer under the direction of the Responsible Engineer. This is primarily a visual review that considers the actual condition of each installation in a systematic manner. It is generally referred to as a "walkdown" or "walkthrough" review because the engineers performing the review systematically walk down each equipment item, building, or system to look for potential seismic vulnerabilities. The basis for assessment may include observed failure modes from past earthquake experience, basic engineering principles, and engineering judgment. The walkdown review emphasizes the primary seismic load resisting elements and the potential areas of weakness due to design, construction, or modification practices, as well as deterioration or damage. A special emphasis is placed on details that may have been designed without consideration of seismic loads. Specific guidance for ground supported tanks is discussed in Section 6. Specific guidance for piping systems is discussed in Section 7.

In many cases, the walkdown review should be supplemented by a review of related drawings. This may be done, for example, to check adequacy of older reinforced concrete structures, to verify anchorage details, or to identify configurations that cannot be visually reviewed due to obstructions, fireproofing, insulation, etc. Note that drawings may not always be available, in which case the engineer should document assumptions made and the basis for those assumptions.

The walkdown review is also used to identify whether or not calculations are needed to complete the evaluation and for what items. The amount of calculations will depend on several factors including the experience of the reviewer, the size, age and condition of the facility, the type of construction, etc. The engineer may choose to evaluate several "bounding cases" or "questionable items" and use those as a basis for further assessments. The calculations should use the guidelines in Section 4 or other appropriate methods.

A detailed description of the walkdown process can be found in ASCE guidelines (Reference 6). Examples of walkdown evaluation sheets are provided in Figure 6.1 of Reference 6 for equipment and References 7 and 8 for piping (see Attachment B). Items of concern identified in the walkdown should be addressed in the seismic report.

4.0 EVALUATION OF GROUND SUPPORTED BUILDING AND NONBUILDING STRUCTURES

4.1 Ground Motion – Define ground motion and response spectra as outlined in Section 2.

4.2 Analysis Methodology and Acceptance Criteria – Acceptance for existing ground supported building and nonbuilding structures (including pressure vessels), and their foundations may be accomplished by one of the following methods. Analysis methods described below may also be used in Sections 5 through 7.

4.2.1 Linear Static and Linear Dynamic Analyses – Perform an appropriate linear dynamic analysis or equivalent static analysis.

The evaluation consists of demonstrating that capacity exceeds demand for identified systems. Acceptance is presumed if the following equation is satisfied:

<table>
<tr><td>DEMAND*</td><td></td><td>CAPACITY BASED ON</td></tr>
<tr><td>$$D + L + \dfrac{E_e}{Q}$$</td><td>$\leq$</td><td>$\emptyset R_n$</td></tr>
</table>

*using Load Factors of unity for all loads
Where,
D = Dead load
L = Live and/or operating load
E_e = Unreduced elastic earthquake load based upon ground motion determined in Section 2
Q = Ductility based reduction factor per Table 1
$\emptyset$ = Strength reduction factor (per ACI) or resistance factor (per AISC)
R_n = Nominal strength (per ACI) or nominal resistance (per AISC)

And subject to the following considerations:

1) For systems whose fundamental period (T) is less than the period at which the peak spectral acceleration occurs (Tpeak), one of the following approaches should be used to determine the appropriate level of seismic acceleration for the fundamental and higher modes. [Note: Tpeak is the period at which the ground motion has the greatest spectral amplification. For spectra that have flattened peaks (e.g., ASCE/SEI 7-05 Figure 11.4-1), the smallest period of the flattened peak (T_0) should be used.]

   a. The peak spectral acceleration should be used for the fundamental mode of the structure. When considering higher modes, either the peak or actual spectral acceleration values may be used.

   b. For a structure that has a fundamental period less than $0.67 \times T_{peak}$, the maximum spectral acceleration in the range of $0.5 \times T$ to $1.5 \times T$ may be used in lieu of the peak spectral acceleration. When considering higher modes, either the peak or actual spectral acceleration values may be used.

2) For redundant structural systems, (e.g., multiple frames or multiple bracing systems), in which seismic loads can be redistributed without failure, the demand (from the previous equation) on an individual frame or member may exceed its capacity by up to 50 percent, provided that the structure remains stable. In addition, the total seismic demand on the structure should not exceed the capacity of the overall structure.

3) Relative displacements should be considered and should include torsional and translational deformations. Structural displacements that are determined from an elastic analysis that was based on seismic loading reduced by Q, should be multiplied by the factor Q to determine displacements to be used in an evaluation.

   a. Generally, the drift (relative horizontal displacement) should be less than 0.02 H, where H is the height between levels of consideration. This drift limit may be exceeded if it can be demonstrated that greater drift can be tolerated by structural and nonstructural components or the equipment itself.

   b. To obtain relative displacements between different support points, absolute summation of the individual displacements can conservatively be used. Alternatively, the Square Root of the Sum of Squares (SRSS) method for combining displacements may be used where appropriate.

4) The potential for overturning and sliding of the foundation should be evaluated. When evaluating overturning, a minimum of 10 percent reduction in dead load should be assumed to account for vertical acceleration effects. This reduction factor may be higher for facilities close to active faults that may be subject to higher vertical acceleration. The factor of safety against overturning and sliding should be larger than or equal to 1.0, considering the appropriate Q-factor from Table 1.F.

5) The capacity of existing concrete anchorage may be evaluated in accordance with the strength design provisions of Section 1923 of the 1997 UBC with inspection load factors specified in Section 1923.3 taken as unity. Alternatively, the capacity of existing concrete anchorage may be evaluated in accordance with the strength provisions of ACI 318-11 Appendix D excluding the requirements of D.3.3.4.3.(a).3, D.3.3.4.3.(d) and D.3.3.5.3.(c).

6) The directional effects of an earthquake should be considered either using the Square Root of the Sum of the Square (SRSS) rule or the 100%-30%-30% rule.

7) Structures that do not pass these evaluation criteria can be reassessed using a more rigorous approach to determine if structural retrofit is actually required.

8) Note that the importance factor (I), as defined in the ASCE/SEI 7-10 (Reference 4) base shear equation for design of new facilities, should be set to unity (1.0) for evaluation of existing facilities, unless an importance factor greater than 1.0 is requested by the owner of the facility.

9) For soil bearing and piping and pressure vessel designs where working stress allowable design is standard practice, capacity may be taken as 1.6 times working stress allowable (without the 1/3 increase).

4.2.2 Nonlinear Static and Nonlinear Dynamic Analyses – Alternative procedures using rational analyses based on well established principles of mechanics may be used in lieu of those prescribed in these recommendations. Methods such as nonlinear time history and nonlinear static pushover analyses would be acceptable. The resulting inelastic deformations should be within appropriate levels to provide reasonable assurance of structural integrity. Acceptable methods include those provided in ASCE/SEI 7-10 (Reference 4) Section 16.2 or ASCE 41-13 (Reference 25). For significant structures, where these types of analyses are preferred, a peer review should be done.

4.2.3 Recommended Guidelines for Seismic Evaluation and Design of Petrochemical Facilities – ASCE (Reference 6), Section 4.0, including appendices, provides a summary of analytical approaches as well as detailed examples for the evaluation of structural period, base shear and other pertinent topics.

5.0 EVALUATION OF EQUIPMENT AND NONSTRUCTURAL COMPONENTS

Permanent equipment and nonstructural components supported within or by structures as indicated in Section 1.2 should be assessed together with the supporting structure. If the equipment or component is directly founded on soil or ground, it should be treated separately as a nonbuilding structure per Section 4.

The supported permanent equipment and nonstructural components should be considered subsystems if their total weight is less than 25% of the total weight of the supporting structure and subsystems. For these subsystems, the anchorage and attachments may be evaluated in accordance with the equivalent static force provisions of Chapter 13 of ASCE/SEI 7-10. The equipment or the nonstructural component itself should be checked for the acceleration levels based on the above referenced sections. Alternatively, a modal dynamic analysis using the evaluation basis spectra as defined in Section 2 of this document, may be performed in accordance with equation 13.3-4 of ASCE/SEI 7-10 if the equivalent static force provisions of Chapter 13 result in excessive demand. Also, nonlinear dynamic analysis is permitted of combined nonstructural systems in accordance with Section 4.2.2.

If the permanent equipment or nonstructural component weight is greater than 25% of the weight of the supporting structure, Section 4 with Q values equal to the smaller of the values for the equipment or the supporting structure from Table 1 can be used for the entire system. Alternatively, a dynamic analysis of the equipment coupled with the supporting structure may be performed to determine the elastic response of the equipment. The elastic responses should then be reduced by the smaller Q value to obtain the design values.

Where an approved national standard provides a basis for the earthquake-resistant design of a particular type of nonbuilding structure, such a standard may be used, provided the ground motion used for analysis is in conformance with the provisions of Section 2.

6.0 EVALUATION OF GROUND SUPPORTED STORAGE TANKS

6.1 Scope – Vertical liquid storage tanks (commonly called flat bottom storage tanks) with supported bottoms at ground level should be addressed using the approaches provided in this section when they meet one of the criteria in Section 1.2. These are tanks which either (a) contain an RS, (b) contain fluids (firewater being the most common example) which are required in an emergency, or (c) are located sufficiently close to a tank in one of the two previous categories so as to pose a threat to the covered process or its emergency shutdown. Horizontal vessels (bullets), vertical vessels and spherical tanks which are supported at ground level are addressed in Section 4.0. Elevated tanks and vessels are addressed in Section 5.0.

Section 7.0 of Reference 6 provides a thorough overview of tank failure modes during a seismic event, seismic vulnerabilities to look for during a seismic walkdown, and the detailed methodology for analytical evaluation as well as suggested modifications to mitigate seismic hazards. See Figure 7.7 of that document for valuable illustrations of some of the items of concern, which typically include over-constrained piping, stairway and walkway attachments to the tank.

6.2 Tank Damage in Past Earthquakes – Vertical liquid storage tanks with supported bottoms have often failed, sometimes with loss of contents during strong ground shaking. The response of such tanks, unanchored tanks in particular, is highly nonlinear and much more complex than that generally implied in available design standards. The effect of ground shaking is to generate an overturning force on the tank, which in turn causes a portion of the tank bottom plate to lift up from the foundation. While uplift, in and of itself, may not cause serious damage, it can be accompanied by large deformations and major changes in the tank shell stresses. It can also lead to damage and/or rupture of the tank shell at its connection with any attachments (e.g., piping, ladders, etc.) that are over-constrained and cannot accommodate the resulting uplift. Tanks have been observed to uplift by more than 12 inches in past earthquakes.

The following are typical of the failure (or damage) modes of tanks that have been observed during past earthquakes:

1) Buckling of the tank shell known as "elephant foot" buckling. This typically occurs near grade around the perimeter of unanchored tanks. Another less common (and less damaging) buckling mode of the tank shell, normally associated with taller tanks, is "diamond shape" buckling.

2) Weld failure between the bottom plate and the tank shell as a result of high-tension forces during uplift.

3) Fluid sloshing, thus potentially causing damage to the tank's roof and/or top shell course followed by spillage of fluid.

4) Buckling of support columns for fixed roof tanks.

5) Breakage of piping connected to the tank shell or bottom plate primarily due to lack of flexibility in the piping to accommodate the resulting uplift.

6) Tearing of tank shell or bottom plate due to over-constrained stairway, ladder, or piping anchored at a foundation and at the tank shell. Tearing of tank shell due to over-constrained walkways connecting two tanks experiencing differential movement.

7) Non-ductile anchorage connection details (anchored tanks) leading to tearing of the tank shell or failure of the anchorage.

8) Splitting and leakage of tank shells due to high tensile hoop stress in bolted or riveted tanks.

6.3 Recommended Steps for Tank Evaluation – When evaluating existing ground supported tanks for seismic vulnerabilities, the following steps should be followed:

1) Quantification of site-specific seismic hazard as outlined in Section 2.

2) Walkdown inspection to assess piping, staircase and walkway attachments, and other potential hazards.

3) Analytical assessment of tanks to evaluate the potential for overturning and shell buckling. Such analysis may usually be limited to tanks having a height-to- diameter ratio of greater than 0.33.

Engineering judgment of the evaluating engineer should be relied upon to determine the need for analytical evaluations. Considerations such as presence of ductile anchorage, plate thickness, favorable aspect ratio of the tank, operating height, ductile tank material, weld/bolting detail, etc. are important in determining whether an analytical assessment is required. Two evaluation methods are provided below in Sections 6.3.1 and 6.3.2.

6.3.1 Linear Static Analysis of Tanks - Linear static analysis procedures are provided in the following industry standards. These include:

1) API 650 Appendix E (Reference 9) - This method is a standard for the design of new tanks for the petrochemical industry. Its provisions are accepted by the CBC and ASCE/SEI 7-10 and it addresses both anchored and unanchored tanks.

2) AWWA D100 (Reference 10) - This method is very similar to the API 650 method and is used primarily for design of water storage tanks. It addresses both anchored and unanchored tanks.

3) Veletsos and Yang (Reference 11) - This method is primarily for anchored tanks.

4) Manos (Reference 12) - This method was primarily developed to evaluate the stability of unanchored tanks and is based on correlation between empirical design approach and observed performance of tanks during past earthquakes. It is generally less conservative than API 650.

5) Housner and Haroun (Reference 13) - This method is primarily for the analysis of anchored tanks, but is often used for both anchored and unanchored tanks.

6) ACI 350.3-01, (Reference 14) - Applies to Concrete Tanks (both round and rectangular)

7) API 620 Appendix L (Reference 26)

8) "Nuclear Reactors and Earthquakes". United States Atomic Energy Commission, TID-7024, August 1961 (Reference 27)

Alternatively, the Q factor given in Table 1 for tanks in conjunction with the demand equation in Section 4.2.1 may be used to determine the lateral seismic loads for tanks. As a guidance, the Q factor method may be used for non-metallic as well as smaller less significant tanks whereas the more traditional methods in the literature as listed above may be used for larger tanks (metallic and concrete). It should be noted that in References 9 and 10 listed above, Q factor reductions are inherently included in the determination of seismic forces. In References 11 to 14 listed above, the Q factors should only be applied to impulsive or structural modes (not sloshing modes).

6.3.2 Nonlinear Static Analysis of Tanks - Section 4.2.2 allows that nonlinear static analysis is an alternative procedure that can be used to evaluate existing structures. Although there are no published guidelines on how to apply this methodology to bottom-supported liquid storage tanks, the following is a suggested approach that can be deemed as acceptable if other methods do not result in demonstrating adequate seismic resistance.

A vertical liquid storage tank may be evaluated using a nonlinear static analysis procedure such as the following:

The loading should be composed of both static fluid pressures, which are constant, plus the effects of fluid inertia forces which are simulated by monotonically increasing two pressure profiles on the tank walls and bottom. The fluid inertia force profiles may be taken from Appendix F of TID 7024 (Reference 27), which contains the original derivation of seismic-induced fluid inertial forces as derived by Housner. The two pressure profiles are (a) those for the portion of the fluid which moves with the tank (termed the impulsive portion), and (b) those for the portion of the fluid which "sloshes" (termed the convective portion). Both portions contain horizontal pressure profiles on the sides of the tank and a vertical pressure profile on the tank bottom.

The pressure profiles are to be monotonically increased until a horizontal "target displacement" for the design earthquake is exceeded at the maximum fluid level. The target displacement may be calculated using Equation 3-14 of ASCE 41-13 (Reference 25). When using this empirical equation for the calculation of the target displacement, in lieu of specific data, the product of the three "C" coefficients need not exceed 1.5.

For thin walled tanks, diamond and elephant foot buckling are potential limit states which can be evaluated by using either recognized equations for storage tank wall stress state at incipient buckling (Reference 29 and 30) or by detailed nonlinear finite element analysis. The analysis is typically a nonlinear pushover analysis where the fluid inertial loads are increased until a post peak in the load-displacement curve is observed.

The acceptance criteria for the seismic-resisting elements of the tank, including anchor bolts and foundation, should be as follows. For deformation-controlled elements (as defined in ASCE 41-13), the plastic deformation of these elements should not exceed deformations consistent with a "collapse prevention" level of performance. For force-controlled elements (again as defined in ASCE 41-13), the seismic force in the specific element at target displacement may be reduced by a "Q" factor as per Section 4.2.1 of this document. However, for such force-controlled elements (such as shell buckling and anchor bolts whose ultimate load is governed by concrete failure), the "Q" factor should not exceed 2.5.

6.4 Mitigation Measures for Tanks – If the walkdown and the evaluation of the tank identify potential seismic vulnerabilities, mitigation measures should be considered. These mitigations may include measures such as increasing the tank wall section (e.g., ribs), addition of flexibility to rigid attachments, reduction of safe operating height or, as a last resort, anchorage of the tank.

6.5 Sloshing Effects – The height of the convective (sloshing) wave (d_s) may be calculated by the following equation:

$$d_s = 0.42 D_i S_a$$

Where,

D_i = the diameter of a circular tank, or the longer plan dimension of a rectangular tank.

S_a = the spectral acceleration, as a fraction of g, at the convective (sloshing) period.

The period (T) of the convective (sloshing) mode in a circular tank may be calculated by the following formula:

$$T = 2\pi \sqrt{\frac{D_i}{3.68 \cdot g \cdot \tanh\left(\frac{3.68 \cdot H}{D_i}\right)}}$$

Where,

H = the height of the fluid,

g = the acceleration due to gravity in consistent units

The above equation for amplitude of a sloshing wave is appropriate for fixed roof tanks. However, in lieu of a detailed analysis, the above equation may be used for a floating roof tank if the weight of the floating roof is replaced by an equivalent height of fluid.

1) For fixed roof tanks, the effects of sloshing may be addressed by having sufficient freeboard to accommodate the wave slosh height. However, when this is not possible, then the following steps should be incorporated into the tank evaluation (or the design of mitigation measures): The geometry of the wave (both unconfined and confined by the roof) should be defined. The geometry of the unconfined wave may conveniently be taken as a trapezoid or a parabola.

2) The fluid head of the freeboard deficit (the unconfined wave height less the available freeboard) should be considered to act as an upward load on the roof. The roof live load should not be considered as assisting to resist this upward fluid pressure.

3) The mass of the fluid that is in the sloshing wave but within the portion confined by the roof should be considered to act laterally at the period of the structural (or impulsive) mode, rather than at the period of the sloshing mode.

For floating roof tanks, the key concern is that the slosh height will be sufficient to lift the bottom of the floating roof onto the top of the shell, potentially leading to a release of contents. Since most tank shells cannot sustain such a weight, this could also result in a major risk of buckling or other failure of the shell at the top of the shell.

It should be noted that the T_L value in the seismic hazard formulation defines the long period response that affects sloshing. The engineer should be aware that significant sloshing can occur even at low seismicity sites. There are numerous documented instances of sloshing related damage at sites over 100 miles from the epicenter that had negligible short period ground shaking.

7.0 EVALUATION OF PIPING SYSTEMS

7.1 Aboveground Piping Systems – Evaluation of piping systems should be primarily accomplished by field walkdowns. One reason this method is recommended is because some piping is field routed and, in some instances, piping and supports have been modified from that shown on design drawings.

The procedure for evaluating aboveground piping systems should be as follows:

1) Identify piping systems to be evaluated. The list should include piping systems that can directly, or indirectly, lead to a significant release of RS as discussed in Section 1.2. The list should also include piping downstream of relief valves and other safety systems used to remove RS to a safe location.

2) Perform a walkdown of the piping systems for seismic capability. Document the walkdown and identify areas for detailed evaluation, if any.

3) Complete the detailed evaluation of any identified areas and recommend remedial actions, if required.

Damage to or failure of pipe supports should not be construed as a piping failure unless it directly contributes to a pressure boundary failure. The intention here is to preserve the essential pressure containing integrity of the piping system but not necessarily leak tightness. Therefore, this procedure does not preclude the possibility of small leaks at bolted flange joints.

The guidance provided in Sections 7.1.1 through 7.1.6 is primarily intended for ductile steel pipe constructed to a national standard such as the American Society of Mechanical Engineers (ASME) B31.3 (Reference 15). Evaluation of other piping material is discussed in Section 7.1.7. The basis for certain provisions in this section and further discussion can be found in Reference 8.

7.1.1 Historical Piping Earthquake Performance – Ductile piping systems have, in general, performed adequately in past earthquakes. Where damage has occurred, it has been related to the following aspects of piping systems:

1) Excessive seismic anchor movement. Seismic anchor movements could be the result of relative displacements between points of support/attachment of the piping systems. Such movements include relative displacements between vessels, pipe supports, or main headers for branch lines.

2) Interaction with other elements. Interaction is defined as the seismically induced impact of piping systems with adjacent structures, systems, or components, including the effects of falling hazards.

3) Extensive corrosion effects. Corrosion could result in a weakened pipe cross section that could fail during an earthquake.

4) Non-ductile materials such as cast iron, fiberglass, glass, etc., combined with high stress or impact conditions.

7.1.2 Walkdown – The walkdown is the essential element for seismic evaluations of piping systems. Careful consideration needs to be given to how the piping system will behave during a seismic event, how nearby items will behave during a seismic event (if they can interact with the piping system) and how the seismic capacity will change over time. The walkdown should be performed in accordance with Section 3. Some guidance on how to perform a walkdown can be found in Reference 6.

Additional aspects of piping systems which should also be reviewed during the walkdown for seismic capability are:

1) Large unsupported segment of pipe (see ASME B31E (Reference 21) Table 2)
2) Brittle elements
3) Threaded connections, flange joints, and special fittings
4) Inadequate supports, where an entire system or portion of piping may lose its primary support

Special features or conditions to illustrate the above concerns include:

1) Inadequate anchorage of attached equipment
2) Short/rigid spans that cannot accommodate the relative displacement of the supports (e.g., piping spanning between two structural systems)
3) Damaged supports including corrosion
4) Long vertical runs subject to inter level drift
5) Large unsupported masses (e.g., valves) attached to the pipe
6) Flanged and threaded connections in high stress locations
7) Existing leakage locations (flanges, threads, valves, welds)
8) Significant external corrosion
9) Corrosion Under Insulation (CUI)
10) Inadequate vertical supports and/or insufficient lateral restraints
11) Welded attachments to thin wall pipe
12) Excessive seismic displacements of expansion joints
13) Brittle elements such as cast iron pipes
14) Sensitive equipment impact (e.g., control valves)
15) Potential for fatigue of short to medium length rod hangers that are restrained against rotation at the support end

7.1.3 Analysis Considerations – Detailed analysis of piping systems should not be the focus of this evaluation. Rather it should be on finding and strengthening weak elements. However, after the walkdown is performed and if an analysis is deemed necessary, the procedures in ASME B31E (Reference 21) and the following general rules should be followed.

1) Friction resistance should not be considered for seismic restraint, except for the following condition: for long straight piping runs with numerous supports, friction in the axial direction may be considered

2) Spring supports (constant or variable) should not be considered as seismic supports

3) Unbraced pipelines with short rod hangers can be considered as effective lateral supports if justified

4) Appropriate stress intensification factors ("i" factors) should be used

5) Allowable piping stresses should be reduced to account for fatigue effects due to significant cyclic operational loading conditions. In this case the allowables presented in Section 7.1.7 may need to be reduced.

6) Flange connections should be checked to ensure that high moments do not result in significant leakage

7.1.4 Seismic Anchor Movement – The recommended procedure for seismic anchor movement (SAM) evaluation of piping is discussed in Section 3.4 of ASME B31E (Reference 21) including allowable stress values. The relative seismic anchor displacements should be calculated following the methodology in Section 4.2.1.

7.1.5 Interaction Evaluation – The recommended procedures for interaction evaluation of piping are as follows:

1) RS piping should be visually inspected to identify potential interactions with adjacent structures, systems, or components. Those interactions which could cause unacceptable damage to piping, piping components (e.g., control valves), or adjacent critical items should be mitigated.
Note that restricting piping seismic movement to preclude interaction may lead to excessive restraint of thermal expansion or inhibit other necessary operational flexibility.

2) The walkdown should also identify the potential for interaction between adjacent structures, systems or components, and the RS piping being investigated. Those interactions that could cause unacceptable damage to RS piping should be mitigated. Note that falling hazards should be considered in this evaluation.

3) Displacements used when considering seismic interaction should be those calculated per Section 4.2.1 (3).

7.1.6 Inertia Evaluation – The recommended procedure for seismic inertia evaluation of piping is discussed in Section 3.4 of ASME B31E (Reference 21) including allowable stress values. Seismic loading should be determined following Section 13.3.1 of Reference 4. The value of Rp should be substituted with Q from Table 1 and the value of Ip=1.0.

7.1.7 Allowable Stress – Piping made from materials other than ductile steel accepted by ASME B31 may be required to withstand seismic loading. The criteria outlined above for ductile steel piping should be followed for piping made from other materials with the following allowable stress values:

1) When ductile material piping is designed and constructed to a national standard with basic allowable stresses given, then those values should be used multiplied by the appropriate factor in Section 3.4 of B31E.

2) When piping materials meet a national standard with a minimum specified tensile strength, σ_t, then the basic allowable stress at operating temperature should be:

   a. Ductile Materials: $S_h = \sigma_t / 3$ at temperature

   b. Brittle Materials: $S_h = \sigma_t / 10$ at temperature

3) When piping materials cannot be identified with a national standard with a minimum specified tensile strength, then one should be estimated from published literature or a testing program. The basic allowable stress at temperature should be determined using the appropriate equation in (2) above, unless a higher allowable can be justified by seismic testing.

7.2 Underground Piping Systems – Piping that is underground should be identified as such on walkdown reports and other documentation prepared for this evaluation. The evaluator can use the technical guidance provided in the aboveground piping section or other technical guidance appropriate for underground piping seismic evaluations. Concerns unique to underground piping that should be considered by the engineer include:

1) Liquefaction and lateral spreading

2) Seismic settlement

3) Surface faulting

Additional evaluation guidance for underground piping systems can be found in Reference 28.

8.0 STRENGTHENING CRITERIA

A strengthening and/or management program should be developed to correct deficiencies. If strengthening is required, appropriate strengthening criteria should be developed to provide a confidence level that retrofitted items will perform adequately when subjected to strong earthquake ground motions.

An important point to consider when retrofitting is that over-strengthening areas of the structure that are currently deficient in strength can force the weak link(s) to occur in other elements that are perhaps more brittle. This can have a negative impact on overall structural performance during a major earthquake. In other words, a structure that is presently weak, but ductile, should not be strengthened to the point that its failure mode becomes brittle with a lower energy absorbing capacity.

Often, the largest category of structural/seismic deficiencies in an existing facility will involve equipment which is not anchored or braced and thus has no lateral restraint. This may include equipment or structures for which bracing has been omitted or removed, or it may include structural bolts or anchor bolts, including their nuts, which were never installed. Another deficiency might be structural elements that are severely corroded or damaged. For such items, the strengthening measures may be obvious, or at least straightforward.

For "building-like" nonbuilding structures (those with framing systems that are specifically listed in the building codes), the procedures and analysis methods outlined in documents such as ASCE 41-13 (Reference 25) may be useful in determining appropriate strengthening measures.

When seismic hazards such as liquefaction or seismically induced landslide can potentially affect a site, it is recommended that a geotechnical engineer be consulted. The basic reference for assessing these seismic hazards is SP117 (Reference 5). However, Section 12 of Reference 16, developed by the Los Angeles Section of ASCE, gives additional guidelines for mitigating landslide hazards. Section 8 of Reference 17, also developed by the Los Angeles Section of ASCE, gives additional guidelines for mitigating liquefaction hazards at a site.

When any retrofit construction work associated with the CalARP program is to be undertaken, a Building Permit is normally required; thus the local Building Department is involved automatically. It should always be kept in mind that the intent of retrofitting these structures, systems, or components is not "to bring them up to current code." In many instances,"to bring them up to current code" may not be practical. The retrofit design criteria should be consistent with this proposed guidance. However, it is always advisable to meet code requirements to the extent practical. If the retrofit construction does not meet the current Building Code, the detail drawings should clearly state that the retrofit is a voluntary seismic upgrade and may not meet current Building Code requirements for new construction.

The concept of "grandfathering" of existing structures is addressed specifically in Sections 3403 to 3405 of the 2013 CBC Those sections of the code basically set out conditions whereby the entire structure need not be brought up to current code when additions, alterations or repairs are made. In addition to requiring that the newly designed portion itself meet the current code, the primary requirements for "grandfathering" the unaltered portion of the structure are that the change cannot increase the gravity load in existing elements by more than 5% without meeting new code requirements for the gravity loads, and that the seismic demand-capacity ratios (DCRs) in existing elements cannot increase by more than 10% without meeting new code requirements. Note that the basis of comparison is the structure with the alteration versus the structure with the alteration ignored. The original design code is not relevant. Additional conditions are provided in Section 3404 of the 2013 CBC and restrictions for "voluntary seismic improvements" are provided in Section 3404.5. The consensus of the Committee is that allowing this type of "grandfathering" of existing structures is appropriate.

If the intent of any retrofit construction associated with the CalARP programs is to do enough work to satisfy the CalARP Program requirements (to mitigate the perceived risk of an accidental release of the regulated substance), but not meet the current code requirements. It behooves the owner and/or the Responsible Engineer, as appropriate to discuss the proposed work with the local Building Code Official to ensure the Building Code Official is in agreement.

9.0 RECOMMENDED REPORT CONTENTS

The CalARP seismic assessment report should contain the items listed below as applicable. There are two types of reports which can be generated. The first type is the initial, or first time, report for a system which has no prior CalARP seismic evaluation. The second type is the revalidation report which is for a system that has one or more prior CalARP seismic evaluations.

9.1 Initial Report Contents – CalARP seismic reports should at least contain the following information:

1) Provide a statement that this is an initial CalARP seismic evaluation report.

2) Provide a description of the scope of the structural/seismic evaluation as determined in Section 1.2. This description may be in terms of the RS present at the facility and where in the facility those RS are located (area, building, floor, etc.). The scope description should include a listing or a tabulation of the items in the facility that was reviewed including structures, equipment and piping. Key items which were specifically excluded and therefore were not reviewed should also be noted.

3) Provide a characterization of the soil profile at the site, and the basis for that characterization. For example, reference to a geotechnical report (e.g., see Attachment A for recommended contents of a Geotechnical report), including its date of issue, if such report serves as the basis for the site soil profile characterization (as per the guidelines in Section 2). In addition, if the geotechnical report serves as the basis for assessing the potential for any of the seismic hazards in Section 2, this should be noted. Depending on the extent to which the geotechnical report is relied upon, it may be appropriate to append a copy of this geotechnical report, or at least key excerpts from it, to the CalARP seismic report.

4) Provide a discussion of the determination of each of the seismic hazards listed in Section 2, and the basis for the determination of each. In particular, where ground response spectra are used as the basis for the CalARP seismic assessment, they should be referenced along with the basis for determining the ground response spectra (See Section 2.1).

5) For each reviewed item, provide an assessment of its structural adequacy to resist the estimated seismic ground shaking for the site.

   a. The assessment should include a noting of any deterioration in the physical condition of the reviewed item that was observed in the field walkdown, such as excessive corrosion, concrete spalling, etc.

   b. The assessment should indicate the basis used. This would include visual observations made during a walkdown and corroborating photographs. Depending on the circumstances, the assessment may also be based on drawing reviews and structural/seismic calculations.

6) Provide recommendations for conceptual measures that will alleviate seismic deficiencies. These recommendations may include:

    a. Strengthening of structural elements

    b. Addition of new structural elements

    c. Reduction or redistribution of the seismic forces

    d. Measures for reducing the effects of a seismic hazard as identified in Section 2, etc.

7) Provide a recommendation for further study or detailed design for items that appear to be seismically deficient or for items which are clearly deficient but for which an adequate seismic risk-reduction measure is not obvious. Such further study may involve a structural issue or it may involve a study on how to address a seismic hazard in Section 2.

8) The initial CalARP report should be signed and stamped by the Responsible Engineer (see Section 1.5).

9) The CalARP report should discuss all deficiencies and recommendations identified during this evaluation. Provide a photograph showing the identified deficiency if possible.

10) A list of the drawings that were reviewed should be included (including date and revision number) when drawing reviews form part of the basis for determining the seismic adequacy of structures or equipment.

11) Supplementary documentation of the observations made and the assessments performed. These may include photographs (where permissible) and copies of walkdown sheets.

9.2 Revalidation Report Contents – CalARP seismic reports should at least contain the following information:

1) Provide a statement that this is a revalidation CalARP seismic report.

2) Provide a description of the scope of the structural/seismic evaluation as determined in Section 1.2. This description may be in terms of the RS present at the facility and where in the facility those RS are located (area, building, floor, etc.). The scope description should include a listing or a tabulation of the items in the facility that is to be reviewed including structures, equipment and piping. Key items which were specifically excluded and therefore were not reviewed should also be noted.

3) If available, provide an updated characterization of the soil profile at the site, and the basis for that characterization. For example, reference to a geotechnical report (e.g., see Attachment A for recommended contents of a Geotechnical report), including its date of issue, if such report serves as the basis for the site soil profile characterization (as per the guidelines in Section 2). In addition, if the geotechnical report serves as the basis for assessing the potential for any of the seismic hazards in Section 2, this

should be noted. Depending on the extent to which the geotechnical report is relied upon, it may be appropriate to append a copy of this geotechnical report, or at least key excerpts from it, to the CalARP seismic report.

4) Provide a discussion of the determination of each of the seismic hazards listed in Section 2, and the basis for the determination of each. In particular, where ground response spectra are used as the basis for the CalARP seismic assessment, they should be referenced along with the basis for determining the ground response spectra (See Section 2.1). Compare the current CalARP seismic loading to the prior evaluations seismic loading and comment.

5) Provide a discussion of items with a recommendation for remediation or additional evaluation from a prior evaluation and list the status of these prior recommendations. Prior recommendations should be categorized as having been sufficiently addressed, partially addressed with further action still required, or not addressed. Prior recommendations should always be completed unless the reviewer can demonstrate in writing that the prior recommendation is not needed anymore and the basis for this determination.

6) For each reviewed item, provide an assessment of its structural adequacy to resist the estimated seismic ground shaking for the site.

   a. The assessment should include a noting of any deterioration in the physical condition of the reviewed item that was observed in the field walkdown, such as excessive corrosion, concrete spalling, etc.

   b. The assessment should indicate the basis used. This would include visual observations made during a walkdown and corroborating photographs. Depending on the circumstances, the assessment may also be based on drawing reviews or structural/seismic calculations.

7) Provide recommendations for conceptual measures that will alleviate seismic deficiencies. These recommendations may include:

   a. Strengthening of structural elements

   b. Addition of new structural elements

   c. Reduction or redistribution of the seismic forces

   d. Measures for reducing the effects of a seismic hazard as identified in Section 2, etc.

8) Provide a recommendation for further study or detailed design for items that appear to be seismically deficient or for items which are clearly deficient but for which an adequate seismic risk-reduction measure is not obvious. Such further study may involve a structural issue or it may involve a study on how to address a seismic hazard in Section 2. Include prior recommendations that were not addressed or which were not addressed adequately since the last evaluation.

9) The revalidation CalARP report should be signed and stamped by the Responsible Engineer (see Section 1.5).

10) The revalidation CalARP report should discuss all deficiencies and recommendations identified during this evaluation regardless of whether or not they were contained in previous evaluation findings. Provide a photograph showing the identified deficiency if possible.

11) A list of the drawings that were reviewed should be included (including date and revision number) when drawing reviews form part of the basis for determining the seismic adequacy of structures or equipment.

12) Supplementary documentation of the observations made and the assessments performed. These may include photographs (where permissible) and copies of walkdown sheets.

10.0 REFERENCES

References may be obtained from:

Engineering Societies Library (Linda Hall Library), a private library located on the campus of the University of Missouri
5109 Cherry Street
Kansas City, Missouri 64110-2498
1-800-662-1545

1. California Code of Regulations (CCR) Title 19, Division 2 Chapter 4.5, *California Accidental Release Prevention (CalARP) Program Detailed Analysis.*
2. *Guidance for California Accidental Release Prevention (CalARP) Program Seismic Assessments*, Prepared for the Administering Agency (AA) Subcommittee Region I Local Emergency Planning Committee (LEPC), Prepared by the CalARP Program Seismic Guidance Committee, September 2010, Approved by Region I LEPC, 1/14/2010.
3. *Uniform Building Code*, 1997 Edition, International Conference of Building Officials, Whittier, California, 1997.
4. ASCE/SEI 7-10, *Minimum Design Loads for Buildings and Other Structures*, American Society of Civil Engineers, Reston, Virginia, 2010.
5. California Department of Conservation, *California Geological Survey, Guidelines for Evaluating and Mitigating Seismic Hazards in California*, Special Publication 117A, 2008.
6. *Guidelines for Seismic Evaluation and Design of Petrochemical Facilities*, Second Edition, Task Committee on Seismic Evaluation and Design of Petrochemical Facilities, American Society of Civil Engineers, Reston, Virginia, 2011.
7. Gurbuz, O., A. Lopez, and P. Summers, April 1992, "Implementation of the Proposed RMPP Seismic Assessment Guidance to Perform a Structural Seismic Evaluation of Existing Facilities" in *Proceedings of HAZMACON '92*, Session on "New Developments in Earthquake Caused Hazardous Materials Releases," Long Beach, CA.
8. Saunders, K.L., and G. Hau "Seismic Evaluation Acceptance Guideline for Existing Above Ground Piping Systems", ASME PVP Volume 256-1, 1993.
9. API Standard 650, *Welded Tanks for Oil Storage*, 12th Edition, American Petroleum Institute, Washington, D.C., 2013.
10. AWWA D100-11, *Welded Carbon Steel Tanks for Water Storage*, American Water Works Association, Denver, Colorado, 2011.
11. Veletsos, A.S., Contributor, "Guidelines for the Seismic Design of Oil and Gas Pipeline Systems", ASCE, Committee on Gas and Liquid Fuel Lifelines, NY, NY, 1984.
12. Manos, G.W., August 1986, "Earthquake Tank-Wall Stability of Unbraced Tanks," American Society of Civil Engineers, Journal of Structural Engineering, Vol. 112, No. 8, including Erratum in Journal of Structural Engineering, Vol. 113, No.3, March 1987.
13. Housner, G.W., and M.A. Haroun, 1980, "Seismic Design of Liquid Storage Tanks," ASCE Convention and Exposition, Portland, Oregon, April 14-18, 1980.
14. ACI 350.3-06, *Seismic Design of Liquid - Containing Concrete Structures and Commentary*, American Concrete Institute, Farmington Hills, Michigan, 2006.
15. ASME B31.3 - 2012, *Process Piping, ASME Code for Pressure Piping, B31*, American Society of Mechanical Engineers, New York, New York, 2013.
16. ASCE, Los Angeles Section Geotechnical Group, "Recommended Procedures for Implementation of Division of Mines and Geology Special Publication 117, Guidelines

for Analyzing and Mitigating Landslide Hazards in California", Published by Southern California Earthquake Center (SCEC), June 2002.

17. ASCE, Los Angeles Section Geotechnical Group, "Recommended Procedures for Implementation of Division of Mines and Geology Special Publication 117, Guidelines for Analyzing and Mitigating Liquefaction in California", Published by Southern California Earthquake Center (SCEC), March 1999.

18. FEMA 350 *Recommended Seismic Design Criteria for New Steel Moment-Frame Buildings*, Federal Emergency Management Agency, Washington, D.C., June 2000.

19. 2012 *International Building Code*, International Code Council, Inc., Country Club Hills, Illinois, June 2011.

20. FEMA P646 *Guidelines for Design of Structures for Vertical Evacuation from Tsunamis*, Federal Emergency Management Agency, Washington, D.C., June 2008.

21. ASME B31Ea – 2010, *Addenda to ASME B31E – 2008 Standard for the Seismic Design and Retrofit of Above – Ground Piping Systems*, American Society of Mechanical Engineers, New York, New York, 2010.

22. *2013 California Building Code Title 24, Part 2, Volumes 1 and 2*, California Building Standards Commission, Sacramento, California, 2013.

23. National Science and Technology Council, *Tsunami Risk Reduction for the United States: A Framework for Action*, December 2005.

24. Bernard, B, L Dengler, and S Yim (editors), 2007, *National Tsunami Research Plan: Report of a Workshop Sponsored by NSF/NOAA.*

25. ASCE/SEI 41-13, *Seismic Rehabilitation of Existing Buildings*, American Society of Civil Engineers, Reston, Virginia, 2013.

26. API Standard 620, *Design and Construction of Large, Welded, Low-pressure Storage Tanks*, Eleventh Edition, American Petroleum Institute, Washington, D.C., 2008.

27. US Department of Energy, *Nuclear Reactors and Earthquakes*, Report No. TID-7024, January 1961.

28. *Guidelines for the Design of Buried Steel Pipe*, American Lifelines Alliance, July 2001.

29. M.J.N Priestley (Editor & Chairman) et. al, "Seismic Design of Storage Tanks," Recommendations of a Study Group of the New Zealand National Society for Earthquake Engineering, December 1986.

30. Eurocode 8, (1998), "Design provisions for earthquake resistance of structures, Part 1-General rules and Part 4 - Silos, tanks and pipelines," European committee for Standardization, Brussels.

31. *CSU Seismic Requirements*, The California State University, Office of the Chancellor, December 21, 2011.

Table 1. Ductility-Based Reduction Factors (Q) for Existing Structures and Systems

A. STRUCTURES SUPPORTING EQUIPMENT   **Q**

This covers structures whose primary purpose is to
support equipment, such as air coolers, spheres,
horizontal vessels, exchangers, heaters, vertical vessels
and reactors, etc.

1. Steel structures

Ductile moment frame (see Note 8)   6 or 8

Use Q=6 if there is a significant departure from the
intent of the 1988 (or later) UBC for special moment-
resisting frames.

Ordinary moment frame (see Note 8)   2, 4 or 5

The following structural characteristics are usually
indicative of a Q=2 value (also see Note 7):

a. There is a significant strength discontinuity in any of
   the vertical lateral force resisting elements, i.e., a
   weak story.

b. There are partial penetration welded splices in the
   columns of the moment resisting frames.

c. The structure exhibits "strong girder-weak column"
   behavior, i.e., under combined lateral and vertical
   loading, hinges occur in a significant number of
   columns before occurring in the beams.

The following structural characteristics are usually
indicative of a Q=4 value (also see Note 7):

d. Any of the moment frame elements is not compact.

e. Any of the beam-column connections in the lateral
   force resisting moment frames does not have both:
   (1) full penetration flange welds; and (2) a bolted or
   welded web connection.

f. There are bolted splices in the columns of the
   moment resisting frames that do not connect both
   flanges and the web.

(Continued)

Table 1. Ductility-Based Reduction Factors (Q) for Existing Structures and Systems (Continued)

Braced frame	2, 4 or 5

The following structural characteristics are usually indicative of a Q=2 value (also see Note 7):

a. There is a significant strength discontinuity in any of the vertical lateral force resisting elements, i.e., a weak story (see ASCE7-10 Table 12.3-2).

b. The bracing system includes "K" braced bays. Note: "K" bracing is permitted for frames of two stories or less by using Q=2. For frames of more than two stories, "K" bracing must be justified on a case-by-case basis.

c. Brace connections are not able to develop the capacity of the diagonals.

d. Column splice details cannot develop the column capacity.

The following structural characteristics are usually indicative of a Q=4 value (also see Note 7):

e. Diagonal elements designed to carry compression have (kl/r) greater than 120.

f. The bracing system includes chevron ("V" or inverted "V") bracing that was designed to carry gravity load and/or beams not designed to resist unbalanced load effects due to compression buckling and brace yielding.

g. Tension rod bracing with connections which develop rod strength.

Cantilever column	2 or 3.5

The following structural characteristics are usually indicative of a Q=2.0 value (also see Note 7):

a. Column splice details cannot develop the column capacity.

b. Axial load demand represents more than 20% of the axial load capacity.

(Continued)

Table 1. Ductility-Based Reduction Factors (Q) for Existing Structures and Systems (Continued)

	Q
A. STRUCTURES SUPPORTING EQUIPMENT (Continued)	
2. Concrete structures	
Ductile moment frame	6 or 8

Use Q=6 if there is a significant departure from the intent of the 1988 (or later) UBC for special moment-resisting frames. If shear failure occurs before flexural failure in either beam or column, the frame should be considered an ordinary moment frame.

Intermediate moment frame   4 1.5, 2.5 or 3.5

Ordinary moment frame

The following structural characteristics are usually indicative of a Q=1.5 value (also see Note 7):

   a. There is a significant strength discontinuity in any of the vertical lateral force resisting elements, i.e., a weak story.

   b. The structure exhibits "strong girder - weak column" behavior, i.e., under combined lateral and vertical loading, hinges occur in a significant number of columns before occurring in the beams.

   c. There is visible deterioration of concrete or reinforcing steel in any of the frame elements, and this damage may lead to a brittle failure mode.

   d. Shear failure occurs before flexural failure in a significant number of the columns.

The following structural characteristics are usually indicative of a Q=2.5 value (also see Note 7):

   e. The lateral resisting frames include prestressed (pretensioned or post-tensioned elements).

   f. The beam stirrups and column ties are not anchored into the member cores with hooks of 135° or more.

   g. Columns have ties spaced at greater than d/4 throughout their length. Beam stirrups are spaced at greater than d/2.

   h. Any column bar lap splice is less than 35 d_b long. Any column bar lap splice is not enclosed by ties spaced 8 d_b or less.

   i. Development length for longitudinal bars is less than 24 d_b.

   j. Shear failure occurs before flexural failure in a significant number of the beams.

(Continued)

Table 1. Ductility-Based Reduction Factors (Q) for Existing Structures and Systems (Continued)

Shear wall   1.5, 3 or 5

The following structural characteristics are usually indicative of a Q=1.5 value (also see Note 7):

a. There is visible deterioration of concrete or reinforcing steel in any of the frame elements, and this damage may lead to a brittle failure mode.

b. There is a significant strength discontinuity in any of the vertical lateral force resisting elements, i.e., a weak story.

c. Any wall is not continuous to the foundation. The following structural characteristics are usually indicative of a Q=3 value (also see Note 7):

d. The reinforcing steel for concrete walls is not greater than 0.0025 times the gross area of the wall along both the longitudinal and transverse axes. The spacing of reinforcing steel along either axis exceeds 18 inches.

e. For shear walls with H/D greater than 2.0, the boundary elements are not confined with either: (1) spirals; or (2) ties at spacing of less than 8 d_b.

f. For coupled shear wall buildings, stirrups in any coupling beam are spaced at greater than 8 d_b or are not anchored into the core with hooks of 135° or more.

(Continued)

Table 1. Ductility-Based Reduction Factors (Q) for Existing Structures and Systems (Continued)

Cantilever pier/column	1.5, 2.5 or 3.5

The following structural characteristics are usually indicative of a Q=1.5 value (also see Note 7):

a. There is visible deterioration of concrete or reinforcing steel in any of the elements, and this damage may lead to a brittle failure mode.

b. Axial load demand represents more than 20% of the axial load capacity.

The following structural characteristics are usually indicative of a Q=2.5 value (also see Note 7):

c. The ties are not anchored into the member cores with hooks of 135° or more.

d. Columns have ties spaced at greater than d/4 throughout their length. Piers have ties spaced at greater than d/2 throughout their length.

e. Any pier/column bar lap splice is less than 35 d_b long. Any pier/column bar lap splice is not enclosed by ties spaced 8 d_b or less.

f. Development length for longitudinal bars is less than 24 d_b.

B. EQUIPMENT BEHAVING AS STRUCTURES WITH INTEGRAL SUPPORTS	**Q**
1. Vertical vessels/heaters or spheres supported by:	
Steel skirts	2 or 4

The following structural characteristics are usually indicative of a Q=2 value (also see Note 7):

a. The diameter (D) divided by the thickness (t) of the skirt is greater than 0.441*E/F_y, where E and F_y are the Young's modulus and yield stress of the skirt, respectively.

(Continued)

Table 1. Ductility-Based Reduction Factors (Q) for Existing Structures and Systems (Continued)

Steel braced legs without top girder or stiffener ring	1.5, 3 or 4

The following structural characteristics are usually indicative of a Q=1.5 value (also see Note 7):

a. The bracing system includes "K" braced bays.
b. Brace connections are not able to develop the capacity of the diagonals.
c. Column splice details cannot develop the column capacity.

The following structural characteristics are usually indicative of a Q=3 value (also see Note 7):

d. Diagonal elements designed to carry compression have (kl/r) greater than 120.
e. The bracing system includes chevron ("V" or inverted "V") bracing that was designed to carry gravity load and/or beams not designed to resist unbalanced load effects due to compression buckling and brace yielding.
f. Tension rod bracing with connections which develop rod strength.

Steel unbraced legs without top girder or stiffener ring	1.5 or 2.5

The following structural characteristics are usually indicative of a Q=1.5 value (also see Note 7):

a. Column splice details cannot develop the column capacity.
b. Axial load demand represents more than 20% of the axial load capacity.

2. Chimneys or stacks	
Steel guyed	4
Steel cantilever	4
Concrete	4

(Continued)

Table 1. Ductility-Based Reduction Factors (Q) for Existing Structures and Systems (Continued)

	Q
C. PIPEWAYS	**Q**
Notes:	

1) This includes pipeways supporting equipment that does not weigh more than 25% of the other dead loads. For pipeways supporting equipment that weighs more than 25% of the other dead loads, see Section A, STRUCTURES SUPPORTING EQUIPMENT.
2) In order to use the full Q-values below, the caveats mentioned earlier in Section A of this table for steel and concrete structures must not be present. If they are, the Q-values in Section A should be used.

	Q
1. Steel	
Ductile moment frame (see Note 8)	8
Ordinary moment frame (see Note 8)	6
Braced frame	6
Cantilever column	4
2. Concrete	
Ductile moment frame	8
Ordinary moment frame	5
Cantilever column	3.5
D. GROUND SUPPORTED TANKS	**Q**
(see Notes 4 and 9)	
1. Anchored	4
2. Unanchored	3
E. FOUNDATIONS (See Note 5)	**Q**
1. Piled	6
2. Spread footings	6
F. ANCHORAGE TO CONCRETE (see Note 6 and 9)	**Q**
1. Anchorage in tension and/or shear when there is a ductile force transfer mechanism between structure and foundation.	As for structure
2. Anchorage in tension and/or shear when there is a non-ductile force transfer mechanism between structure and foundation.	1.5

(Continued)

Table 1. Ductility-Based Reduction Factors (Q) for Existing Structures and Systems (Continued)

	Q
G. PIPING	**Q**
1. Piping in accordance with ASME B31, including in-line components with joints made by welding or brazing.	4.5
2. Piping in accordance with ASME B31, including in-line components, constructed of high- or limited-deformability materials, with joints made by threading, bonding, compression couplings, grooved couplings or flanges.	4
3. Piping and tubing not in accordance with ASME B31, including in-line components, constructed of high-deformability materials, with joints made by welding or brazing.	4
4. Piping and tubing not in accordance with ASME B31, including in-line components, constructed of high- or limited-deformability materials, with joints made by threading, bonding, compression couplings, grooved couplings or flanges.	3.5
5. Piping and tubing constructed of low-deformability materials, such as cast iron, glass, and nonductile plastics.	3

NOTES:

1. The use of the highest Q-factors in each category requires that the elements of the primary load path of the lateral force resisting system have been proportioned to assure ductile rather than brittle system behavior. This can be demonstrated by showing that each connection in the primary load path has an ultimate strength of at least equal to 150% of the load capacity (governed by either yielding or stability) of the element to which the load is transferred. Alternatively, Q-factors should be reduced consistent with the limited ductility of the governing connection and/or the governing connection should be modified as required.

2. A Q-factor different from the tabulated values (higher or lower) may be justified on a case-by-case basis.

3. If more than one of the conditions specified in the table applies, the lowest Q-factor associated with those conditions should be used.

4. Other approved national standards for the seismic assessment of tanks may be used in lieu of these guidelines.

5. These values of Q apply to overturning checks, soil bearing, and pile capacities. For the remaining items including connection between piles and pile caps, use the Q factor for the supported structure.

6. For anchorage in tension or shear a ductile force transfer mechanism can be defined as when the concrete-governed strength is greater than 1.2 times the anchorage steel strength or when there are properly detailed concrete reinforcing bars being provided that prevent a concrete failure. When this is the case, then the Q-factor to be used for the evaluation of the anchorage, and the rest of the structural system corresponds to that for the structural system itself. For the evaluation and design of concrete reinforcing bars, a ductility factor of 0.75Q of the structure or Q=1.5, whichever is greater, shall be used. Additionally, for vessels, stacks and tanks the use of properly proportioned anchor bolt chairs must also be provided to ensure a ductile force transfer.

For anchorage in tension or shear a non-ductile force transfer mechanism can be defined as when a concrete-governed strength controls the evaluation of anchorage (as opposed to anchorage steel and in situations where inadequate reinforcing is provided) or when there is some other non-ductile force transfer mechanism between the structure and foundation. When this is the case, then a Q-factor of 1.5 should be used for the evaluation of the anchorage.

Combined tension and shear interaction shall also be checked for both the steel anchorage itself and the concrete embedment. If either tension or shear reinforcing is provided to prevent concrete failure then no interaction effects need to be considered for the concrete strength evaluation.

Additionally, for skirt supported vessels, flat bottom tanks or other structural systems where the anchorage is the primary source for ductility, the Q-factor determined for the anchorage shall also be used for the evaluation of vessel or tank itself or structural system. Also see Note 7.

Where anchorage corrosion is found then the effective area of the anchorage shall be reduced accordingly and taken into account in determining the anchorage strength. If the anchorage corrosion is severe enough to prevent adequate ductile yielding of the anchorage then a Q-factor of 1.5 shall be used for the anchorage evaluation.

7. Alternatively, for structures that may contain localized/single features with limited ductility, such as limiting connections or splices, non-compact steel members, high (Kl/r) members and non-ductile anchor bolts, that do not occur at a significant number of locations, the load capacity of the specific limiting feature(s) may be evaluated and/or improved in lieu of using system-wide lower Q-factors that tend to generically penalize all elements of the structural system. The evaluation for these localized features may be performed using a Q-factor equal to 0.4 times the Q-factor normally recommended (i.e., unreduced) for the system. The evaluation for the remainder of the system may then be performed using the Q-factor normally recommended without consideration of the localized feature with limited ductility

8. Figure 1 below shows a common connection detail which has been used in the building industry. In the aftermath of the January 1994 Northridge, California earthquake, over 100 buildings were found, where cracks occurred in connections based on this detail. This Committee suggests that for determining the connection forces using a Q-value equal to one half (1/2) of Q for the structure system, but not less than 2, where this type of connection is present, unless justified otherwise.

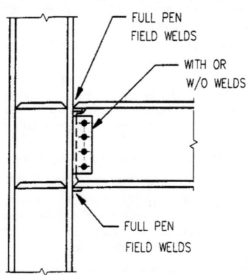

FULL PEN
FIELD WELDS

WITH OR
W/O WELDS

FULL PEN
FIELD WELDS

Figure 1: Former Standard Ductile Moment Connection Detail. (As a result of the Northridge Earthquake, this connection was shown to have major problems.)

9. For tanks made of fiberglass or similar materials, non-ductile anchorage and its attachments should be evaluated for a Q equal to 1.5.

ATTACHMENT A RECOMMENDED GEOTECHNICAL REPORT CONTENTS

A proper assessment of the above earthquake hazard effects will generally require, as a prerequisite, knowledge of the underlying soil profile at the facility. Therefore, a geotechnical report for the facility should be made available to the engineer performing the CalARP seismic review.

If the soil profile is known to be uniform over the entire area, a geotechnical report developed for an adjacent facility may be adequate. It is preferable if the adjacent site having a geotechnical report is within 300 feet of the facility in question. Consultation with the AA and with the local Building Official may also provide some information in this regard.

If the owner cannot provide an adequate geotechnical report, then the options are as follows:

1) The owner may contract with a licensed geotechnical engineer to provide a report that will be adequate for the CalARP seismic review.

2) The engineer may engage a licensed geotechnical engineer as a sub-consultant to provide a geotechnical report.

3) The engineer may make a series of conservative (essentially "worst case") assumptions in determining the effects of the underlying soil profile on the various seismic hazards. Such assumptions may be based on the soil characteristics known for the general area. Alternatively, the site class may be assumed which gives the largest evaluation forces. Depending on the situation, this option may or may not be the most cost-effective approach for the owner (e.g., for a single small item, it is generally not cost effective to have a geotech report performed).

A standard geotechnical investigation report should include the information in the following list. The listed items are divided into two "tiers" or types of information. The first tier lists the basic minimum contents of a geotechnical (soils) report. The second tier lists information which the engineer performing the CalARP seismic review will eventually require, and it will be convenient and beneficial if the geotechnical report provides a professional presentation of this information.

Tier 1 – Minimum Contents of Geotechnical Report for CalARP Review

1) Plot Plans drawn to scale depicting the locations of exploratory borings.

2) Boring logs (to depth of at least 50 feet) indicating ground surface elevation, blow counts (penetration), graphic log of material encountered, depth to groundwater (if encountered), soil classification and description (per ASTM standards), moisture content and dry density.

3) Geologic setting, subsurface soil conditions soil types, and regional ground-water information.

4) Recommendations for appropriate foundation schemes and design parameters including soil bearing capacity, estimated total/differential settlements, and lateral resistance.

5) Recommendations for the design of retaining walls including active and passive earth pressures.

Tier 2 – Desirable Additional Contents of Geotechnical Report

1) Recommendations pertaining to seismic design parameters based on ASCE/SEI 7-05 or the latest California Building Code adopted by the local jurisdiction. Parameters such as Site Class; Site MCE Ground Motion Parameters S_S and S_1, Site Coefficients F_a and F_v and site DE parameters S_{DS}, S_{D1} and T_L.

2) Results of geologic and seismic hazard analysis (based on guidelines in SP117) including poor soil conditions, locations of active and potentially active faults, fault rupture potential, liquefaction, seismically-induced settlement/differential settlements, and seismically-induced flooding.

ATTACHMENT B WALKDOWN FORMS FOR EQUIPMENT AND PIPING

FIELD DATA SHEET FOR EQUIPMENT
EQUIPMENT ID: **DESCRIPTION:** **LOCATION:**
SCREENING EVALUATION: SUMMARY
Summary of Evaluation: _____ Adequate _____ Not Adequate _____ Further Evaluation Required **Recommendations:**
SCREENING EVALUATION: ANCHORAGE
Noted Anchorage Concerns: _____ Installation Adequacy   _____ Weld Quality _____ Missing or Loose Bolts   _____ Corrosion _____ Concrete Quality   _____ Other Concerns _____ Spacing/Edge Distance **Comments:**
SCREENING EVALUATION: LOAD PATH
Noted Load Path Concerns: _____ Connections to Components   _____ Missing or Loose Hardware _____ Support Members   _____ Other Concerns **Comments:**

Seismic Evaluation
CalARP Walkdown Review Sheet
Piping

Line Number:		Date:	
Drawing Number:		By:	

Evaluation Summary (Circle one)

Adequate   Not Adequate   Further Evaluation Required

Inspection Attributes				
	Yes	No	Inac	Comments
Piping				
Damaged				
Corrosion				
Flanged/Threaded Joints				
Buried Run				
Adequate Branch Flexibility				
Rigidly Spans Components				
Supports				
Piping Spans OK				
Missing Hardware				
Corrosion				
Hardware Damaged/Loose				
Seismic Interaction				
Adequate Clearance				
Adjacent Comps. Secure				
Clearance at AOVs/MOVs				

Line Number:		Date:	

Notes and Sketches

APPENDIX 4.G EXAMPLES OF CONFIGURATIONS OF PETROCHEMICAL STRUCTURES WHERE DYNAMIC ANALYSIS IS RECOMMENDED

This appendix provides guidelines for choosing the type of seismic analysis procedure to apply to petrochemical structures, assuming that the least complicated type of analysis is the first choice of analysis type. This does not preclude that a more complicated type of analysis cannot and should not be used. For instance, structures in Seismic Design Categories A through C do not require dynamic analysis; therefore, the recommended seismic analysis procedure is the equivalent lateral force analysis. Dynamic analysis procedures, however, may be used for structures in Seismic Design Categories A through C and may yield an overall reduction in seismic load effects compared with those calculated by equivalent lateral force analysis. The engineer must carefully choose the type of seismic analysis for a structure based upon all contributing factors, including but not limited to seismic design category, structural irregularities, fundamental natural period, structure height, and economics.

Dynamic analysis is not required for nonbuilding structures similar to buildings if they meet any one of the following criteria:

- Structures in Seismic Design Categories A through C;
- Risk Category I or II structures not exceeding about 24 ft in height;
- Structures of light framed construction;
- Structures with no structural irregularities and not exceeding 160 ft in structural height;
- Structures exceeding 160 ft in structural height with no structural irregularities and with $T > 3.5T_s$; and
- Structures not exceeding 160 ft in height and having only horizontal irregularities of type 2, 3, 4, or 5 of ASCE 7 Table 12.3-1, or vertical irregularities of type 4, 5a, or 5b of ASCE 7 Table 12.3-2.

ASCE 7 permits the use of equivalent lateral force analysis for any nonbuilding structure not similar to buildings regardless of the seismic design category, risk category, fundamental natural period, height, or structural irregularities. However, for some structural configurations dynamic analysis is recommended. Table 4.G-1 provides guidelines for some structural configurations where dynamic analysis is recommended. Dynamic analysis should be performed by a competent analyst and should include a peer review. In cases where the participation of higher modes will significantly influence the seismic load effect, E, dynamic analysis should be performed.

Table 4.G-1. Structural Configurations Where Dynamic Analysis Is Recommended.

Item	Sketch of Nonbuilding Structure	Description of Nonbuilding Structure (Including Criteria)
1		Flexible nonbuilding structure, in which W_p is greater than $0.25(W_s + W_p)$, that is supported on a relatively flexible elevated support structure. Flexibility of the attachment and supports should be considered.
2		Nonbuilding braced frame structures that provide nonuniform horizontal support to the equipment. Analysis should include coupled model effects.
3		Heaters with flexible stack supported from the heater (e.g., API 560 Type E heaters).
4		Stacked vertical vessels with significant difference in mass distribution ($W1 > 1.5W2$ or $W1 < 0.67W2$).

(Continued)

Table 4.G-1. Structural Configurations Where Dynamic Analysis Is Recommended. (Continued)

Item	Sketch of Nonbuilding Structure	Description of Nonbuilding Structure (Including Criteria)
5	 Elevation	Vertical vessels with large attached vessels, with $W2$ greater than about $0.25(W1 + W2.)$
6	 Elevation	Nonbuilding structures containing lug-supported equipment with $W2 > 0.25(W1 + W2)$. A coupled system with the mass of the equipment and local flexibility of the supports should be considered in the model. This is especially important when the equipment is supported near its center of mass.
7	 Elevation	Flexible equipment connected by large-diameter, thick-walled pipe and supported by a flexible structure. Should be modeled as a coupled system, including the pipe. A coupled response may be more advantageous to piping and nozzle design than treating the structures independently.

(Continued)

Table 4.G-1. Structural Configurations Where Dynamic Analysis Is Recommended. (Continued)

Item	Sketch of Nonbuilding Structure	Description of Nonbuilding Structure (Including Criteria)
8	Elevation	Coker structures: Coke drums supported by concrete table-top structures with structural steel braced derrick structures on top.
9	ΔMAX ΔAVG Plan	Nonbuilding structure with torsional irregularity: rigid or semi-rigid diaphragm where $\Delta_{MAX} > 1.2\ \Delta_{AVG}$.
10	Elevation	Nonbuilding structure with a soft story irregularity.
11	M4 M3 M2 M1	Nonbuilding structure with weight (mass) irregularity: $M3 > 1.5M2$ or $1.5M4$.

(Continued)

Table 4.G-1. Structural Configurations Where Dynamic Analysis Is Recommended. (Continued)

Item	Sketch of Nonbuilding Structure	Description of Nonbuilding Structure (Including Criteria)
12	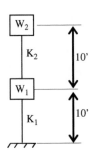 Elevation	Note that ASCE 7 requires a dynamic analysis when the structure has a Vertical Geometric Irregularity (namely, L1 > 1.3L2). This committee, however, feels that dynamic analysis is needed in this case if the stiffness of the two adjacent bays (of length L1 and L2) differ from each other significantly (stiffness of one bay less than 70% of stiffness of the other bay).

Example of Simplified Dynamic Analysis Method

To determine the effect of the vertical structural irregularity on the equivalent static method, consider a problem similar to that in Example, which is now solved using simplifed dynamic analysis based on the Rayleigh method to determine the structural period, mode shape, and equivalent lateral static forces.

From Example 1, the story stiffness of the piperack can be determined as follows:

$$K_2 = 1/(Caa - Cab) = \frac{1}{0.0384 - 0.018} = 49.02 \frac{\text{kip}}{\text{in}}$$

$$K_1 = 1/Cbb = \frac{1}{0.0157} = 63.694 \frac{\text{kip}}{\text{in}}$$

Assume the structure has mass (weight) irregularities such that the effective weight in the upper story is more than 150% of the effective weight in the lower story as follows:

$W_1 = 8$ kip and $W_2 = 20$ kip

The mode shape and the fundamental circular frequency of the system can be initially approximated using the generalized stiffness and mass of the system as follows:

Circular frequency, $\omega = [K^*/M^*]^{0.5}$

where
$M^* =$ generalized mass $= \Sigma M_i^* \phi_i^2$
$K^* =$ generalized stiffness $= F/u_x = g\ [\Sigma M_i^* \phi_i]/u_x$
$u_x =$ deflection at top mass
$\phi_i =$ normalized mode shape

Story	Wi	Mi	Ki	Ui	ϕi	Mi ϕi	Mi ϕi^2
2	20	0.0518	49.02	0.565	1.0000	0.0518	0.0518
1	10	0.0259	63.694	0.157	0.2779	0.0072	0.0020
						0.0590	0.0538

$$M^* = \Sigma Mi^* \phi i^2 = 0.0538 \frac{kip}{in}$$

$$K^* = F/ux = [\Sigma Mi\phi i/ux]g = \frac{0.059 \cdot 386.4}{0.565} = 40.35 \frac{kip}{in}$$

$$\omega = (K^*/M^*)^{0.5} = \sqrt{\frac{40.35}{0.0538}} = 27.386 \frac{rad}{sec}$$

The actual first-mode circular frequency should be fairly close to 27.386 rad/s. However, the assumed mode shape used in calculation may not be the correct one. Note also that the mode shape has little influence on the frequency. The correct mode shape and frequency of each mode can be determined by trial and error as follows:

Step 1. Assume a circular frequency slightly lower than that obtained in the calculation above.

Step 2. Beginning from the top mass with $\phi_2 = 1.0$, determine $M_2\ \omega^2\ \phi_2$, $\Sigma M_i\ \omega^2\ \phi_i$, and $\Sigma M_i\ \omega^2\ \phi_i/K_i$.

Step 3. Calculate the modal displacement at mass 1, $\phi_1 = \phi_2 - \Sigma M_i\ \omega^2\ \phi_i/K_i$.

Step 4. Calculate $M_1\ \omega^2\ \phi_1$, $\Sigma M_i\ \omega^2\ \phi_i$, and $\Sigma M_i\ \omega^2\ \phi_i/K_i$.

Step 5. Calculate the modal displacement at ground 0, $\phi_0 = \phi_1 - \Sigma M_i \, \omega^2 \, \phi_i/K_i$.

Step 1: $\omega = 27\frac{\text{rad}}{\text{sec}}$

Step 2: $M_2 \, \omega^2 \, \phi_2 = 0.0518 \cdot 27^2 \cdot 1.0 = 37.762$
$\quad\quad\;\; \Sigma M_i \, \omega^2 \, \phi_i = 37.76$
$\quad\quad\;\; \Sigma M_i \omega^2_{\;i}/K_2 = \frac{37.762}{49.02} = 0.77$

Step 3. $\phi_1 = \phi_2 - \Sigma M_i \, \omega^2 \, \phi_i/K_i = 1.0 - 0.77 = 0.23$

Step 4. $M_1 \, \omega^2 \, \phi_1 = 0.0259 \cdot 27^2 \cdot 0.23 = 4.343$
$\quad\quad\;\; \Sigma M_i \, \omega^2 \, \phi_i = 37.762 + 4.343 = 42.105$
$\quad\quad\;\; \Sigma M_i \omega^2 \phi_i/K_1 = \frac{42.105}{63.694} = 0.661$

Step 5. $\phi_0 = \phi_1 - \Sigma M_i \, \omega^2 \, \phi_i/K_i = 0.23 - 0.661 = -0.431 < 0$
$\quad\quad\quad\quad\quad\quad\quad\quad\quad\quad\quad\quad\quad$ therefore select other frequency.

The process is then repeated using different ω until the modal displacement at ground $= 0$ at convergence. This trial-and-error procedure could be automated using an Excel spreadsheet calculation as shown below.

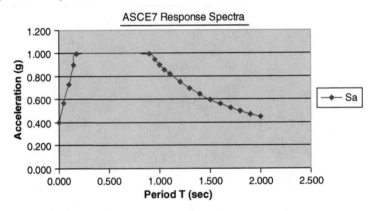

ASCE 7 Response Spectra	
Site =	Class D
Ss =	1.50000
S1 =	0.90000
S_{DS} =	1.00000
S_{D1} =	0.90000
To =	0.18000
Ts =	0.90000

T	Sa
0.000	0.400
0.050	0.567
0.100	0.733
0.150	0.900
0.180	1.000
0.900	1.000
0.950	0.947
1.000	0.900
1.050	0.857
1.100	0.818
1.200	0.750
1.300	0.692
1.400	0.643
1.500	0.600
1.600	0.563
1.700	0.529
1.800	0.500
1.900	0.474
2.000	0.450

STRUCTURAL PERIOD AND MODE SHAPE ANALYSIS
(RAYLEIGH METHOD)

MODE NO.   =   1

TRY PERIOD (T) =   0.2857

ROTATIONAL FREQ. ( ) =   21.988

i	M_i	K_i	i	M_i*	$M_i* *_i$	$(Mi* *_i)$	$(M_i* *_i)/K_i$
2	0.052	49.0	1.0	25.045	25.045	25.045	0.511
1	0.026	63.7	0.5	12.522	6.125	31.169	0.489

i	i		$(Mi*^{2}*i)/Ki$		$(i-1)$
2	1.000	-	0.511	=	0.489
1	0.489	-	0.489	=	0.000

<---- MODE SHAPE AT THE BASE OF STRUCTURE =0.00 AT CONVERGENCE.

MODAL PARTICIPATION FACTOR AND LATERAL FORCE (Fi) DISTRIBUTION:

SPECTRAL ACCELERATION $(Sa=\%g)$ =   1.00

i	M_i	i	M_i*_i	$M_i*_i^{2}$	$F_i=M_{i*} *S_a*386.4*_i$
2	0.052	1.000	0.052	0.052	22.25
1	0.026	0.489	0.013	0.006	5.44
		TOTAL:	0.064	0.058	

1.112

where
i = Story number
M_i = Lumped mass at each story
K_i = Support stiffness for each story
 = Mode shape at each story
F_i = Lateral force at each story = $Mi* *Sa*386.4* i$
 = Mass participation factor ($Mi* i$ $Mi* i^{2}$)

STRUCTURAL PERIOD AND MODE SHAPE ANALYSIS
(RAYLEIGH METHOD)

MODE NO.   =   2

TRY PERIOD (T)   =   0.0905

ROTATIONAL FREQ. ( ) =   69.390

i	M_i	K_i	$_i$	M_i*	$M_i* *_i$	$(M_i* *_i)$	$(M_i* *_i)/K_i$
2	0.052	49.0	1.0	249.416	249.416	249.416	5.088
1	0.026	63.7	-4.1	124.708	-509.811	-260.395	-4.088

i	$_i$		$(M_i*^{2}*_i)/K_i$		(i-1)
2	1.000	-	5.088	=	-4.088
1	-4.088	-	-4.088	=	0.000

<---- MODE SHAPE AT THE BASE
OF STRUCTURE =0.00
AT CONVERGENCE.

MODAL PARTICIPATION FACTO AND LATERAL FORCE (Fi) DISTRIBUTION:

SPECTRAL ACCELERATION ($Sa=\%$g) =   0.70

i	M_i	$_i$	M_i*_i	$M_i*_i^{2}$	$F_i= M_i* *S_a*386.4*_i$
2	0.052	1.000	0.052	0.052	-1.56
1	0.026	-4.088	-0.106	0.433	3.20
		TOTAL:	-0.054	0.485	

-0.112

where:

i = Story number

M_i= Lumped mass at each story

K_i= Support stiffness for each story

$_i$= Mode shape at each story

F_i= Lateral force at each story

 = Mass participation factor ($M_i* i\ M_i* i^{2}$)

Equivalent static lateral forces at each story:

$F2 = [22.25^2 + (-1.56)^2]^{0.5} = 22.305$ kip
$F1 = (5.44^2 + 3.20^2)^{0.5} = 6.311$ kip
Base shear, $V = 22.305 + 6.311 = 28.616$ kip

Now, consider the comparison with equivalent lateral forces determined by using ASCE 7:

$T = 0.29$ s
Base shear, $V = 28 \cdot 1.0 = 28$ kip

$Fx : = \frac{wx \cdot hx^k}{\sum wi \cdot hi^k} \cdot V$ where $k = 1.0$ for T less than or equal to 0.5 s.

$F2 = \frac{20 \cdot 20}{20 \cdot 20 + 8 \cdot 10} \cdot 28 = 23.333$ kip   compare with 22.305 kip

$F1 = \frac{10 \cdot 10}{20 \cdot 20 + 8 \cdot 10} \cdot 28 = 5.833$ kip   compare with 6.311 kip

This comparison shows that the difference in the lateral force determined using dynamic analysis and the equivalent lateral force procedure per ASCE 7 is fairly small; however, the result from this example should not imply that the difference will be small in all cases and that performing dynamic analysis is not needed. This example should be considered as an inital exercise to determine if a more complex computer finite element modeling and dynamic response spectrum analysis is necessary.

CHAPTER 5

Primary Structural Design

5.1 INTRODUCTION

This chapter provides guidance for the seismic structural design of petrochemical facilities. Once the design forces are established using analysis techniques as discussed in Chapter 4, the design may proceed as outlined in this chapter. The methods described are applicable to structures and components, including supports and anchorages of electrical and mechanical systems. Seismic design of electrical and mechanical components (e.g., transformers, pumps, compressors, vessels, etc.) is outside the scope of these guidelines.

The guidelines provided herein are based mainly on current practice and current code provisions. This document was updated to reflect the nonbuilding criteria provided in recent editions of ASCE 7. In addition, criteria and specifications from several petrochemical companies, industrial advisory groups, and architectural and engineering firms were collected and reviewed. The criteria and practices most commonly used were generally preferred and were therefore adopted. Criteria related to special cases may not be universally applicable and therefore were not included. As a result, this chapter includes design recommendations for which a general consensus exists.

5.2 DESIGN CRITERIA

5.2.1 Introduction

Section 5.2 addresses the earthquake load combinations to be used in design and the acceptance criteria to be met. The earthquake load combinations derive mainly from IBC (2018), ASCE 7, and PIP (2017) and are delineated to provide specific guidance for common structures and components at petrochemical facilities. The earthquake load combinations currently used by the petrochemical industry have also been considered. The acceptance criteria are defined by reference to accepted industry codes and standards.

5.2.2 Loads and Earthquake Load Combinations

5.2.2.1 Design Loads

Petrochemical facilities typically have design loads that are unique to the structure or type of equipment. In developing load combinations for seismic design, the engineer should carefully review all applicable loads including, but are not limited to, ice, rain, hydrostatic, dynamic, upset conditions, earth pressure, vehicles, buoyancy, and erection. The design of structures and components should account for the effects of all loads, including those defined as follows:

1. **Dead Loads (D)**

    (a) Dead loads are the actual weight of materials forming the building, structure, foundation, and all permanently attached appurtenances.

    (b) Weights of fixed process equipment and machinery, piping, valves, electrical cable trays, and the contents of these items should be considered as dead loads.

    (c) For this guideline, dead loads are designated by the following nomenclature:

    D_s, D_e, and D_o, where

    D_s = Structure dead load is the weight of materials forming the structure (not the empty weight of process equipment, vessels, tanks, piping, or cable trays), foundation, soil above the foundation resisting uplift, and all permanently attached appurtenances.

    D_e = Empty dead load is the empty weight of process equipment, vessels, tanks, piping, and cable trays.

    D_o = Operating dead load is the empty weight of process equipment, vessels, tanks, piping, and cable trays plus the maximum weight of contents (fluid load) during normal operation. Operating dead load, D_o, consists of dead load, D, and fluid load, F, given in *IBC* nomenclature. Including fluid loads in operating dead loads is an industry standard. Flat bottom tanks are exceptions to this industry standard.

2. **Live Loads (L)**

    (a) Live loads are gravity loads produced by the use and occupancy of the building or structure. These include the weight of all movable loads, such as personnel, tools, miscellaneous equipment, movable partitions, wheel loads, parts of dismantled equipment, stored material, etc.

    (b) Areas specified for maintenance (e.g., heat exchanger tube bundle servicing) should be designed to support the live loads.

    (c) Minimum live loads should be in accordance with ASCE 7 and applicable codes and standards.

3. **Earthquake Loads (E)**

    (a) Earthquake loads include the inertia effects due to a design earthquake. For determination of earthquake loads, refer to Chapter 4.

(b) For the load combinations in Section 5.2, the following designations are used:

E_o = Earthquake load considering the unfactored operating dead load (D_o) and the applicable portion of the unfactored structure dead load (D_s)

Q_{Eo} = Horizontal component of E_o

E_e = Earthquake load considering the unfactored empty dead load (D_e) and the applicable portion of the unfactored structure dead load (D_s)

Q_{Ee} = Horizontal component of E_e

E_{to} = Earthquake load for tanks per API 650, Appendix E (2014)

(c) Vertical component of earthquakes should be considered in all load combinations except for structures assigned to Seismic Design Category B.

The vertical component of earthquakes acting upward need not be considered where determining demands on the soil-structure interface of foundations (e.g., overturning, uplift, and sliding of foundations). For tanks, vessels, hanging structures, and nonbuilding structures incorporating horizontal cantilevers, the vertical seismic term, $0.2S_{DS}$, in the load combinations presented later in this chapter is to be replaced by the term $0.3S_{av}$ determined by ASCE 7-16, Section 11.9. ASCE 7-16, Section 15.1.4 provides additional direction on the use of the $0.3S_{av}$ term..

4. **Thermal Loads**

(a) Thermal loads should be included with operating loads in the appropriate load combinations. Thermal load should have the same load factor as dead load.

(b) Friction loads should be considered temporary and should not be combined with wind or earthquake loads. However, anchor and guide loads (excluding their friction component) should be combined with wind or earthquake loads.

(c) For this guideline, thermal loads are designated by the following nomenclature:

A_f = Pipe anchor and guide forces that are present during normal operation

T_o = Thermal force due to thermal expansion

F_f = Friction force on the sliding heat exchanger or horizontal vessel pier

5. **Pressure Loads (Ground-Supported Tanks Only)**

For this guideline, pressure loads for ground-supported tanks are designated by P_i, P_e, and P_t.

where

P_i = Design internal pressure,

P_e = External pressure, and

P_t = Test internal pressure.

6. **Snow Loads (S)**
Unless otherwise specified, snow loads should be computed and applied in accordance with ASCE 7. Roof live load should be replaced by the snow load when the snow load is greater.

7. **Loads not Combined with Earthquake Loads**
The following loads should not be combined with earthquake loads:

(a) Blast loads: Overpressure due to explosions;

(b) Maintenance loads, including lifting, jacking, and bundle pull loads developed when removing tube bundles from heat exchangers;

(c) Impact loads: Amplified live loads caused by a sudden collision, crash, blow, or drop;

(d) Infrequent traffic loads: Vehicular loads due to periodic use;

(e) Wind loads: Forces cause by atmospheric wind pressure;

(f) Friction loads due to thermal expansion; and

(g) Hydrotest water loads from tank or vessel.

5.2.2.2 Load Combinations

1. **General**
Buildings, structures, equipment, vessels, tanks, and foundations should be designed for the following:

(a) Appropriate load combinations from IBC except as otherwise specified in this Guideline,

(b) Local building codes,

(c) Any other applicable design codes and standards, and

(d) Any other probable and realistic combination of loads.

2. **Typical Earthquake Load Combinations (for Structures and Foundations)**
Load combinations are provided in Tables 5-1 and 5-2 in both allowable stress design (ASD) and strength design format.

Allowable Stress Design

(a) The noncomprehensive list of typical load combinations for each type of petrochemical structure provided in this Guideline should be considered and used as applicable.

(b) Engineering judgment should be used in establishing all appropriate load combinations.

(c) The use of a one-third stress increase for load combinations including wind or earthquake loads should not be allowed for structural steel designs using AISC 360 (2016).

Table 5-1. Load Combinations for Allowable Stress Design.

Load Comb. No.	Load Combination	Description	Notes
1.	$D_s + D_o + A_f + T_o + 0.7E_o$	Operating + Earthquake	
2.	$D_s + D_o + A_f + T_o + 0.75$ $(L + 0.7E_o + S)$	Operating + Live + Earthquake + Snow	Note 9
3.	$0.9D_s + 0.9D_e + 0.7E_e$	Earthquake Uplift (empty)	Note 4
4.	$0.9D_s + 0.9D_o + A_f +$ $T_o + 0.7E_o$	Earthquake Uplift (operating)	Note 4
5.	$D_s + D_o + 0.1S +$ $E_{to} + 0.4P_i$	Operating Weight + Snow + Earthquake + Internal Pressure	Notes 7, 8
6.	$D_s + D_o + 0.1S + E_{to}$	Operating Weight + Snow + Earthquake	Note 8

Note 1: Load combinations 1, 2, 3, and 4 do not apply to atmospheric storage tanks. Load combinations 5 and 6 apply only to atmospheric storage tanks.

Note 2: This is not intended to be a comprehensive list. Some loads may not apply in all cases. The engineer should use judgment to determine a comprehensive list of load combinations that apply to a specific configuration.

Note 3: For skirt-supported vertical vessels and skirt-supported elevated tanks classified as Risk Category IV or designed using an R value of 3 in accordance with ASCE 7, Section 1, the critical earthquake provisions and implied load combination of ASCE 7-16, Section 15.7.10.5, shall be followed.

Note 4: When using load combinations 3 and 4 to evaluate sliding, overturning, and soil bearing at soil-structure interface, do not use the reduction of foundation overturning from ASCE 7-16, Section 12.13.4. Load combinations 4 and 5 are based on the Alternate Basic Load Combinations of IBC Section 1605.3.2 and not ASCE 7, Chapter 2.

Note 5: The design thermal force for horizontal vessels and heat exchangers shall be the lesser of T_o or F_f. If the seismic acceleration is greater than the friction force then this force can usually be neglected.

Note 6: For internal pressures sufficient to lift the tank shell according to the rules of API 650, tank, anchor bolts, and foundation shall be designed to the additional requirements of API 650, Appendix F.7.

Note 7: If the ratio of operating pressure to design pressure of ground-supported storage tanks exceeds 0.4, the owner shall consider specifying a higher factor on design pressure in load combination 5 of Table 5-1.

Note 8: Earthquake loads, E_{to} from API 650 already include the 0.7 ASD seismic load factor.

Note 9: Live loads should not be combined with earthquake loads for extremely low occupancy during normal operation when justified.

Note 10: For seismic loads during construction, see ASCE 37 (ASCE 2014).

(d) The use of a one-third stress increase for load combinations including earthquake loads should not be used for timber and masonry designs using AWC NDS (2018) and TMS 402 (2016), respectively.

Table 5-2. Load Combinations for Strength Design.

Load Comb. No.	Load Combination	Description	Notes
7.	$1.2D_s + 1.2D_o + 1.2A_f + 1.2T_o + f_1L + 1.0E_o + 0.2S$	Operating + Live + Earthquake + Snow	Notes 2, 4, 5
8.	$0.9D_s + 0.9D_e + 1.0E_e$	Earthquake Uplift (empty)	
9.	$0.9D_s + 0.9D_o + 0.9A_f + 0.9T_o + 1.0E_o$	Earthquake Uplift (operating)	

Note 1: This is not intended to be a comprehensive list. Some loads may not apply in all cases. The engineer should use judgment to determine a comprehensive list of load combinations that apply to a specific configuration.

Note 2: If L< 100 psf, then $f_1 = 0.5$. If L >100 psf or areas occupied as places of public assembly, then $f_1 = 1.0$.

Note 3: For skirt-supported vertical vessels and skirt-supported elevated tanks classified as Risk Category IV or designed using an R value of 3 in accordance with ASCE 7, Section 1, the critical earthquake provisions and implied load combination of ASCE 7-16, Section 15.7.10.5, shall be followed.

Note 4: The design thermal force for horizontal vessels and heat exchangers shall be the lesser of T_o or F_f.

Note 5: Live loads should not be combined with earthquake loads for extremely low occupancy during normal operation when justified.

Note 6: For seismic loads during construction, see ASCE 37.

(e) The use of a one-third stress increase for load combinations including earthquake loads may be used for foundation soil bearing and pile capacities if specifically permitted by a registered geotechnical engineer.

(f) Allowable stress load combinations that include earthquake loads should include full snow loads and appropriate live loads when examining maximum load combinations.

Strength Design

(a) The noncomprehensive list of typical factored earthquake load combinations in Table 5-2 should be considered and used as applicable.

(b) Engineering judgment should be used in establishing all appropriate load combinations.

(c) The earthquake load combinations in Table 5-2 are appropriate for use with the strength design provisions of either AISC 360 or ACI 318 (2014).

(d) Snow loads should have a factor, f_2, when combined in strength design load combinations including earthquake loads. Factor f_2 should equal 0.7 for roof configurations (such as saw tooth) that do not shed snow off the structure. Factor f_2 should equal 0.2 for other roof configurations.

Inclusion of snow loads in calculation of seismic loads should be in accordance with Section 4 of this guideline.

(e) Where live loads are present, they should be combined with earthquake loads in strength design load combinations. The inclusion of live loads in the calculation of earthquake loads, however, should be limited to the recommendations of Section 4 of this guideline.

(f) Special seismic load combinations that include the maximum seismic load effect, E_m, should be considered as specified by the IBC and ASCE 7. These load combinations are repeated in Table 5-3.

3. **Load Combinations—Allowable Stress Design**

Table 5-1 should be used for allowable stress design seismic load combinations for buildings, open-framed structures, vertical vessels, horizontal vessels/heat exchangers, pieracks/pipe bridges, and ground-supported tanks.

The use of a one-third stress increase for load combinations including wind or earthquake loads shall not be allowed for designs using AISC 360, AWC NDS, or TMS 402.

4. **Load Combinations—Strength Design**

Table 5-2 should be used for strength design seismic load combinations for buildings, open-framed structures, vertical vessels, horizontal vessels/heat exchangers, and pieracks/pipe bridges.

5. **Load Combinations—Amplified Seismic Strength Design**

Table 5-3 provides the amplified seismic strength design load combinations (refer to ASCE 7-16, Section 12.4.3) that should be used where required for seismic detailing. The $0.2S_{DS}$ listed in the following load combinations should only be applied to vertical dead loads and shall not be applied to horizontal loads. Where amplified load combinations are required for seismic detailing, using amplified strength design load combinations and strength design methods is recommended.

Table 5-3. Load Combinations for Amplified Seismic Strength Design.

Load Comb. No.	Load Combination	Description
10.	$(1.2 + .2S_{DS})(D_s + D_o) + 1.2A_f + 1.2T_o + f_1L + \Omega_oQ_{Eo} + 0.2S$	Operating + Live + Earthquake + Snow
11.	$(1.2 + .2S_{DS})(D_s + D_e) + f_1L + \Omega_oQ_{Ee} + 0.2S$	Empty + Live + Earthquake + Snow
12.	$(0.9 - .2S_{DS})(D_s + D_o) + 0.9A_f + 0.9T_o + \Omega_oQ_{Eo}$	Earthquake Uplift (operating)
13.	$(0.9 - .2S_{DS})(D_s + D_e) + \Omega_oQ_{Ee}$	Earthquake Uplift (empty)

Note: See Table 5-2 notes.

5.2.3 Material Selection

5.2.3.1 General

Proper specification of materials is crucial to the ductile performance of a structure's seismic force–resisting system. In addition, cost and availability must be considered. The following sections provide guidance on the selection of materials specified as acceptable or mandated by the respective code authority.

5.2.3.2 Structural Steel

Although AISC 341 (2016) permits several structural steel specifications, not all shapes are commonly found and availability should be confirmed prior to specifying. Table 5-4 lists some that are commonly specified.

Bolted connections used in seismic load–resisting system joints should be proportioned as pretensioned bearing joints with high-strength bolts as specified in AISC 341. The flaying surfaces of these connections should meet the slip-critical requirements in accordance with AISC 360, Section J3.8, with a Class A surface.

Weld filler used in the members and connections resisting seismic forces should be specified as being capable of producing welds with a minimum Charpy V-Notch toughness of 20 ft-lb (27 J) at 0 °F (−18 °C). Demand-critical welds should be made with a filler metal capable of providing Charpy V-Notch toughness of 20 ft-lb (27 J) at -20_0F (−29 °C) and 40 ft-lb (54 J) at 70 °F (21 °C) when the steel frame is enclosed and normally maintained at a temperature of 50 °F (10 °C) or higher. Service temperatures in structures less than 50°F (10°C) require corresponding reductions in the above qualification temperatures of 20 °F (11 °C) above the lowest anticipated service temperature.

Table 5-4. Commonly Specified Structural Steels for Seismic Resistance.

Shape	ASTM Specification	Comment
W[a]	A992	A36 may be preferred in international locations.
M,S,HP,C,MC,L, Plates and Bars[1]	A36, A572 Grade 50	Check availability of A572. For plates and bars more than 6 in. thick, use A36.
HSS (Round and Rectangular)	A500 Grade B or C, A618	Check availability of A618.
Pipe	A53 Grade B	API 5L Line Pipe PLS 2 Grades are sometimes specified for pipe piling.

[a]Hot-rolled shapes used in the seismic load–resisting system with flanges 1 1/2 in. thick and thicker, plates 2 in. thick and thicker used as built-up members, connection plates where seismic loading inelastic strain is expected, and as the steel core of buckling-restrained braces, should have a minimum Charpy V-Notch toughness of 20 ft-lb (27 J) at 70°F (21°C).

5.2.3.3 Reinforced Concrete

Concrete used in structural members resisting earthquake-induced forces should have a minimum 28-day compressive strength f'_c of 3,000 psi, although 4,000 psi is commonly used for applications in petrochemical facilities. Lightweight concrete is limited to a maximum 28-day f'_c of 5,000 psi. Concrete used in petrochemical applications is frequently subjected to aggressive service due to temperature, machine dynamic loading, and corrosive exposures. Compressive strength is seldom the only important criteria. Extra care should be taken in properly specifying a durable, corrosion-resistant mix design for the application considered so that when seismic demand occurs, the original design cross-sectional properties have not been reduced through the aforementioned aggressive services. The reader is referred to the many American Concrete Institute and Portland Cement Association resources available for more information on concrete mix design.

Reinforcement used in structural frame members and in structural wall boundary elements should comply with ASTM A706. ASTM A615 Grades 60 is allowed, provided that actual mill test results for the material provided show that the yield stress does not exceed the specified f_y by more than 18,000 psi and that the actual tensile strength to actual yield stress ratio is not less than 1.25. When ASTM A615 reinforcing is used under these circumstances, the quality assurance program should address review of the mill-certified material test reports and subsequent matching with material delivered prior to commencing fabrication. Although A706 or A615 with the above limitations is not directly specified for foundations in the seismic provisions of ACI 318, the engineer should consider specifying foundation reinforcement as such to avoid accidental mixing of nonspecification material within flexural frames. A706 reinforcing is also low-alloy with the added benefit of enhanced weldability. It should be specified where carbon steel embed anchorage is provided for by deformed reinforcement or where welding of flexural or compressive reinforcement is required.

5.2.3.4 Masonry and Wood

There are currently no building code prohibitions or restrictions to material specifications used in masonry or wood structures designed to resist seismic forces that are otherwise allowed by code. Specifying quality materials and verifying materials and construction with proper inspection and quality assurance programs are recommended.

5.2.4 Acceptance Criteria

5.2.4.1 General Requirements

Design of petrochemical facilities should comply with the project criteria, which should be developed based on the owner's specifications, design criteria, and applicable standards, some of which are listed in Sections 5.2.4.2 through 5.2.4.5 below.

Local building officials should be consulted to determine codes and standards applicable to the project. The edition date and any addenda and/or supplements should be determined.

5.2.4.2 Steel Acceptance Criteria

The design, quality of steel, fabrication, and erection of steel structures should be in accordance with the requirements of the following references as applicable: IBC, ASCE 7, AISC 360, AISC 341, AISI S100 (2016), ASCE 8 (2002), SJI-100 (2015), and ASCE 19 (2010).

5.2.4.3 Concrete Acceptance Criteria

Design and construction of reinforced concrete structures should conform to the requirements of IBC, ASCE 7, and ACI 318.

5.2.4.4 Masonry

The design, construction, and quality assurance of masonry components that resist seismic forces should conform to the requirements of TMS 402 and TMS 602 (2016), except as modified by the provisions of IBC and ASCE 7.

5.2.4.5 Wood

The design, construction, and quality assurance of members and their fastenings in wood systems that resist seismic forces should conform to the requirements of IBC, ASCE 7, AWC NDS, and AWC SDPWS (2015).

5.3 DESIGN CONSIDERATIONS

5.3.1 Introduction

This section describes typical structures and components found in petrochemical facilities and provides special design recommendations for them.

5.3.2 Buildings

Petrochemical facilities typically contain the following buildings, each with specific function and personnel use requirements:

- Guard or security building,
- Administration building,
- Cafeteria,
- Emergency services (fire/ambulance/medical),
- Laboratory,
- Maintenance shop,

- Warehouse, and

- Control building.

Often two or more of these building functions are combined into a single structure in accordance with client needs or preferences, or for economic advantage. Buildings in petrochemical facilities are typically one or two stories in height.

Structural planning includes evaluating and selecting the building materials, foundation type, and vertical and lateral force-resisting systems for the building. The engineer must exercise care to ensure that codes and standards selected for the design criteria of the building are relevant, comprehensive, nonconflicting, and acceptable to the local building authority.

Seismic design requirements for buildings are contained in Chapter 12 of ASCE 7. Various seismic structural systems and materials of construction may be utilized in the design of petrochemical facility buildings. Construction materials are often determined by economic considerations and local availability. Lateral and vertical force-resisting systems commonly used in petrochemical facility buildings include braced frames, moment-resisting frames, shear walls, or combination systems. Seismic coefficients and detailing provisions for seismic force resisting systems are provided in Table 12.2-1 of ASCE 7-16.

5.3.3 Nonbuilding Structures

5.3.3.1 General

As noted in Chapter 4, nonbuilding structures represent most items in a petrochemical plant. They can be classified as nonbuilding structures with structural systems similar to buildings, or nonbuilding structures not similar to buildings. Chapter 15 of ASCE 7 defines seismic design requirements for nonbuilding structures. The global intent of Chapter 15 is to ensure nonbuilding structures are designed with sufficient strength and ductility to resist seismic design forces in a manner generally consistent with that used for buildings. ASCE 7 recognizes that nonbuilding structures often have unique physical features, which affect their seismic response characteristics. In many cases recognized industry codes and standards govern the design and detailing of these structures. ASCE 7 provides modifications or additions to current versions of industry codes in certain cases to obtain consistency in the seismic design approach.

5.3.3.2 Drift and P-Delta Effects

ASCE 7 does not require nonbuilding structures to meet the drift limitations established for building structures, provided a rational analysis demonstrates that computed drifts will not adversely affect the stability of the structure or the attached components such as piping and walkways. ASCE 7 requires consideration of P-delta effects when critical to the function or stability of the structure. Many engineers prefer to utilize building drift criteria for nonbuilding structures, especially those similar to buildings, expecting these criteria to provide a practical and safe design baseline. In cases where drift limits are exceeded, the engineer then has the choice to strengthen the nonbuilding structure to reduce the drift or further evaluate the computed drift impact on the stability and components.

5.3.3.3 Nonbuilding Structures Similar to Buildings

Piperacks and equipment support structures are the two major categories of nonbuilding structures similar to buildings found in petrochemical facilities. These structures are considered similar to buildings in that they utilize the same vertical and lateral force-resisting systems as buildings do, and they display similar responses to seismic ground motion. Unlike buildings, however, piperacks are open-framed structures, and equipment support structures are typically open-sided or partially sided frames, with or without a roof cover. The key distinction between buildings and nonbuilding structures is that the latter are not designed to be continuously occupied, yet often provide temporary access for operations and maintenance personnel.

Piperacks

Piperacks, also known as pipeways, are open structures in the petrochemical facility that carry piping, conduit, and cable tray within and between process, utility, and offsite areas. Main piperacks are large, multilevel structures routed through the central corridors of the process units and provide interconnection between other process and utility areas. Main piperacks are often used to support finfan coolers above the upper level. Secondary piperacks are smaller, single or multilevel structures carrying piping and commodities to specific equipment or other areas within the facility.

Piperack structural systems are designed as continuous (strutted) or noncontinuous (unstrutted) frames. Resistance to lateral forces in a continuous piperack is provided by a linked series of transverse moment-resisting frames tied into a single braced bay in the longitudinal direction. Expansion joints are used to provide breaks in long runs of continuous piperack. Secondary piperacks may be designed as continuous or noncontinuous frames. Noncontinuous piperacks use freestanding (cantilevered) transverse moment-resisting frames to resist lateral loads in the longitudinal direction.

Piperacks may be constructed of steel, concrete, or a combination thereof. Steel piperacks in process areas are typically fireproofed from grade to a specified elevation for protection against heat damage from a ground-level pool fire. Columns and other gravity load members in piperacks supporting air coolers or equipment containing hydrocarbon materials are fireproofed from grade to the equipment support levels. Common fireproofing materials include concrete, gunite, or lightweight manufactured materials, each having different weight properties and thus affecting the seismic inertia loading.

Equipment Support Structures

Equipment support structures in petrochemical facilities are a broad classification of structures designed for various equipment sizes, weights, and access requirements, typically resulting in unique configurations. Common examples of equipment support structures include heat exchanger/drum support structures, reactor structures, compressor structures, heater structures, and coke drum structures. Equipment support structures include access platforms, ladders, stairs, and monorails or cranes for operations and maintenance.

Vertical bracing or moment-resisting frames provide lateral load resistance in equipment support structures. In steel structures, bracing is commonly used in the weak-axis planes of the columns, but can be used in the strong-axis direction as well. Bracing continuity may be interrupted for personnel access or to avoid interferences with pipe and areas reserved for equipment removal. Moment-resisting frames are also commonly used in the strong-axis direction of the columns, thus allowing more openness within the structure. Beams and columns in steel structures supporting equipment or major pipe systems are typically fireproofed. Concrete is used for equipment support structures where it is economical or to provide superior characteristics for vibration, deflection control, or corrosion resistance.

Seismic Design

Design of the SFRSs for petrochemical piperacks and equipment support structures should comply with the requirements of ASCE 7-16, Table 15.4-1, subject to the system limitations and height limits. This table provides applicable seismic coefficients and references to required design and detailing provisions for nonbuilding structures similar to buildings. Lateral force–resisting systems in common use in petrochemical facility structures are summarized as follows.

Moment-Resisting Frames (Structural Steel and Concrete)

Special moment-resisting frames (SMRFs) allow significant inelastic deformations during a major earthquake and are not generally found in industrial facilities. Instead piperacks and equipment structures typically use intermediate (IMRF) or ordinary (OMRF) moment-resisting frames. IMRFs allow limited inelastic deformations in their members and connections during a major earthquake. These systems allow for seismic load reduction as a function of the response modification factor, R, but require specific levels of ductile behavior. OMRFs are designed to withstand the loading from the design earthquake with minimal inelastic deformation. Connection requirements for OMRFs are prescriptive, whereas those for IMRFs and SMRFs are performance based and must be prequalified or proven by prototype testing.

Braced Frames (Structural Steel)

Concentrically braced frames (X-bracing, V- and inverted V-bracing, and single diagonal bracing) in petrochemical structures can be designed as special concentrically braced frames (SCBF) or ordinary concentrically braced frames (OCBF). SCBFs are expected to withstand significant inelastic deformations when subjected to forces from the design earthquake and therefore have stricter design requirements than OCBFs to ensure ductility and less strength degradation upon buckling of the compression brace. K-bracing is not allowed in SCBFs.

Lateral Deflection

The lateral deflections occurring in piperacks and equipment support structures due to seismic and other design loading must be evaluated to ensure proper performance of the structure. ASCE 7-16, Section 15.5.2.1, provides a formula for computing deflections from seismic loading on piperacks. This formula is also applicable to equipment support structures designed by ASCE 7. Computed lateral seismic deflections should be added to deflections due to other loading in the design combinations, such as the piping anchor forces.

Lateral drift due to earthquake loads is calculated by taking the drift due to the design earthquake load from elastic analysis and multiplying it by a factor of C_d/I_e. This factor accounts for the inelastic component of drift that will occur in ductile lateral force–resisting systems when subjected to the design earthquake loads. Cd is the deflection amplification factor defined in ASCE 7-16, Chapter 11.

Petrochemical structures support equipment and piping containing pressurized hydrocarbon liquid and vapor products. Therefore the engineer must carefully select criteria to ensure that the deflections will not pose an undue risk to safety and property. Permissible lateral deflections for petrochemical piperacks and equipment support structures differ from those intended for buildings. Section 15.4.5 of ASCE 7-16 states that drift limitations for buildings need not apply to nonbuilding structures if a rational analysis indicates they can be exceeded without adversely affecting structural stability or attached or interconnected components and elements such as walkways and piping.

For buildings, setting lateral drift limitations as a percentage of story to story height is practical because of drift limitations that can be sustained by glazing, architectural cladding, sprinkler systems, and other building components. These generalized drift limits do not necessarily apply to petrochemical structures for several reasons. First, petrochemical structures typically do not have drift-sensitive building components. Second, some petrochemical structures support piping systems that are connected to deflection-sensitive nozzles, such as those connected to rotating equipment, and require that the nozzle does not experience large lateral loads during the code-prescribed earthquake loads. Third, some petrochemical structures, such as offsite piperacks, are capable of taking large lateral deflections without damage to the structure or its piping systems. Therefore, building drift limitations may be too strict for offsite piperacks but not strict enough for a structure supporting major piping connected to a compressor.

Coordination between the owner, the engineers who perform pipe stress calculations, the vendors who supply the equipment, and the structural engineer is recommended to develop specific drift criteria for a structure.

5.3.3.4 Nonbuilding Structures Not Similar to Buildings

Petrochemical facilities contain several types of nonbuilding structures not similar to buildings, each with unique seismic design characteristics. These structures include

- Vertical vessels, stacks, elevated tanks, bins, and hoppers (on steel skirt or legs);
- Horizontal vessels or heat exchangers (on saddles or legs);
- Finfans or air-cooled heat exchangers;
- Furnaces, heaters, and boilers;
- Spheres;
- Storage tanks (ground supported);
- Cooling towers; and
- Basins, sumps, and pits.

Chapter 15 of ASCE 7 contains seismic design requirements for nonbuilding structures, both those similar to buildings and those not similar to buildings. Nonbuilding structures may be self-supported or supported by another structure. The method used to determine the design loading for a nonbuilding structures supported by another structure is described in Section 15.3 of ASCE 7-16.

Table 15.4-2 of ASCE 7-16 provides seismic analysis coefficients and required detailing provisions for nonbuilding structures not similar to buildings. Additionally, industry-based consensus standards for seismic and other design aspects for many types of nonbuilding structures are included in ASCE 7-16, Chapter 23.

ASCE 7-16, Section 15.3, describes the approach to analysis for cases where a nonbuilding structure not similar to buildings is supported by or within another structure and is not part of that structure's primary seismic force–resisting system. Petrochemical equipment support structures often fall into the category where the equipment (supported nonbuilding structures) weight is greater than 25% of the combined weight of the structure and equipment. This condition necessitates consideration of the equipment's rigidity in the analysis of the system.

Ground-supported flat bottom tanks are addressed in Chapter 7 of these guidelines.

5.3.4 Elements of Structures, Nonstructural Components, and Equipment

5.3.4.1 General

ASCE 7, Chapter 13 establishes minimum design criteria for a wide range of nonstructural components permanently attached to structures. Chapter 13 segregates nonstructural components and related supports into two primary categories: architectural and mechanical and electrical. Examples of nonstructural components commonly found in petrochemical facilities include

Architectural:

- Building architectural components such as parapets, nonstructural walls and partitions, access floors, and suspended ceilings; and

Mechanical and electrical (located in equipment structures, buildings, or free standing):

- Equipment;
- Process and utility piping systems, including valves and fittings;
- Electrical cable tray, conduit, motor control center (MCC) cabinets, switchgear, panels, and lighting fixtures;
- Control systems cable tray, conduit, and instruments; and
- Building systems including HVAC, lighting, plumbing, and fire protection.

In cases where the weight of a supported component is greater than or equal to 25% of the total weight of the structure and all components, the component itself is considered a nonbuilding structure and designed in accordance with ASCE 7, Chapter 15. For those supported nonbuilding components considered rigid ($T < 0.06$ s for the component + local support structure), the seismic force for the design of the component (F_p) is determined using ASCE 7, Chapter 13, where R_p (component response modification factor) is taken from Table 15.4-2 in ASCE 7-16 and a_p (amplification factor) equals 1.0. For supported nonrigid components greater than or equal to 25% of the total system weight, the seismic design forces are determined by an analysis of the combined structure and components in accordance with ASCE 7-16, Section 15.5. R is taken as the lower value of either the supported nonbuilding structure or the primary supporting structure.

Supported components that represent less than 25% of the system weight include most of the smaller equipment and the piping and electrical systems in petrochemical structures. The seismic design force for these components is determined by the methods provided in Chapter 13. One approach uses values of R_p and a_p from Section 13.1.6 with Equation (13.3-1) to determine F_p. ASCE 7-16, Section C13.3.1, provides a practical method for determining a_p for nonrigid components using the ratio of the component period to the fundamental period of the primary structure. This may allow a_p less than 2.5, resulting in lower seismic design forces. However, for primary structures with a fundamental period greater than T_s (S_{D1}/S_{DS}), higher modes may come into play. Therefore, using a lesser a_p may be unconservative. Alternatively, F_p can be computed using ASCE 7-16, Equation (13.3-4), supported by an appropriate dynamic analysis.

The seismic design force, F_p, is applied in accordance with ASCE 7-16, Section 13.3.1. The code requires accounting for the component force for anchorage of the component to the supporting structure. Although not specifically stated in ASCE 7, a load path evaluation is recommended for important supported components from the point of origin to the foundation of the supporting structure to ensure the safe and efficient transfer of the computed forces. This can be accomplished by specifying a loading combination(s) containing only the seismic component force(s).

5.3.4.2 Typical Supported Equipment

The common permanent equipment supported by structures in petrochemical plants includes the following groups of items.

Cantilever Structures

Cantilever structures include chimneys, stacks, and towers. Design concerns for these items in petrochemical plants are generally no different than supported items in buildings. For these cantilever structures, recognizing that, in many cases, the support level is a flexible structural floor system is especially important. If horizontal deflections near the top of the cantilever item are a concern, increasing the rigidity of the support level to achieve the desired performance may be necessary. Once the loads have been determined and support flexibility issues have been addressed, design procedures for a chimney, stack, tower, or any other similar cantilever structure are the same as if they were supported at grade.

Vessels and Tanks

Vessels and tanks include boilers, heat exchangers, fired and unfired pressure vessels, and miscellaneous tanks. Design concerns for these items in petrochemical plants are generally no different than supported items in buildings. Because these items should be relatively small to be included in this section, support flexibility is relatively unimportant, unless the item is quite long. If the item is long, relative vertical support movement may need to be addressed. Design aspects of vessels and tanks are discussed in Section 5.3.3.4.

Mechanical Equipment

Mechanical equipment includes turbines, chillers, pumps, motors, and air-handling units. Design concerns for these items in petrochemical plants are generally no different than supported items in buildings. These items are usually very rigid compared with the support level. Because these items generally have a low profile, seismic loads rarely, if ever, cause tension forces in the anchorage. However, the connecting structure between the component and the supporting structure is sometimes designed without due consideration for seismic loads. The engineer should review the adequacy of any connecting structures provided by component vendors to ensure that adequate lateral load capacity exists. Anchorage acceptability is evaluated by comparing seismic forces to bolt shear and tension capacity, without frictional resistance from dead load of the item.

Vibration isolators are often utilized with rotating equipment to minimize vibration in the equipment and its supporting structure. The isolators, while solving a difficult vibration problem, may create problems during an earthquake unless care is exercised to restrict horizontal and vertical movement at the isolators. This is often accomplished with snubbers, guides, or stops.

In some cases, mechanical equipment needs to be reviewed. Although specific guidance is not provided herein, seismic design of mechanical equipment should be performed by qualified structural/mechanical engineers. The design should take into consideration inelastic behavior and potential interaction with the supporting structure. ASCE 7 considers these and other factors.

Electrical Equipment and Control Systems

Electrical equipment and control systems include transformers, switchgear, control panels, standby power equipment, and computer equipment. Design concerns for these items in petrochemical plants are generally no different than supported items in buildings. These items are similar to cantilever structures, mechanical equipment, or something in between. Depending on the item's size and shape, follow the appropriate guidance in the paragraphs above on cantilever structures and mechanical equipment.

Distributive Systems

Distributive systems include piping, ducts, conduits, cable trays, and conveyors. Design concerns for these items in petrochemical plants are generally no different than supported items in buildings, except that the pipe sizes in piping systems are usually larger in petrochemical plants. The major design concern for the item itself is the effect of relative movement between points of anchorage.

Leakage at equipment and piping flanges is often due to underdesign of the flange, that is, designing the flange for only operating pressure-temperature conditions. By increasing the flange class, the additional stress resulting from the seismic loads may be accommodated without leakage.

ASCE 7 has specific guidelines that have been developed for the seismic design of distributive systems, such as those noted above. These guidelines include references to codes and standards that are specific to given types of systems. These guidelines are recommended for the design of distributive systems in petrochemical facilities.

5.3.5 Foundations

5.3.5.1 General

Except for certain types of equipment (horizontal vessels, heat exchangers, and vertical vessels), foundations in petrochemical facilities resemble foundations in most other industries. Therefore, the design of foundations should satisfy the requirements of appropriate building code provisions.

Foundations should be designed for seismic loads calculated for the overall structural system. The loads generated by individual elements, for example, brace connections designed for overstrength load, should be considered for anchor bolt design but should not be used for foundation design.

5.3.5.2 Anchorage

Ductile anchorage is required for structures assigned to Seismic Design Categories C, D, E, and F. Designing anchorage for overstrength factor (Ω_o) is not required unless the connection of the anchorage (e.g., structural steel special concentric braced frame connection) is required to be designed for an overstrength factor. For comprehensive seismic design of anchorage see *Anchorage Design for Petrochemical Facilities* (ASCE 2013) and PIP (2018).

5.3.5.3 Overturning

Overturning for load combinations that include earthquake loads should have a factor of safety of 1.0. The factor of safety against overturning should be defined as the summation of moments resisting overturning divided by the summation of overturning moments. The summation of restoring moments should be determined from unfactored service-level loads.

This Guideline recommends using the IBC Alternative ASD Load Combinations when summing overturning moments and restoring moments. The IBC Alternative ASD Uplift Load Combination with earthquake effect includes a $0.9D$ term for dead load resistance and does not include the $0.6D$ term that the ASCE 7 ASD Uplift Load Combination with earthquake effect includes. For this reason, the calculated overturning moment should not be reduced by factors given in ASCE 7-16, Section 12.13.4, for the foundation-overturning checks if IBC Alternative ASD Load Combinations are used.

ACI 318 requires the reinforced concrete of footings to be proportioned to resist factored loads. This requires applying load factors to the loads and calculating a fictitious soil pressure. The reinforced concrete is then proportioned to resist this fictitious soil pressure. When the factor of safety against overturning approaches 1.0, calculating a fictitious soil pressure may not be possible because the eccentricity is outside the footing. In this case, one solution is to substitute a point load at the edge of the footing. The point load can be calculated to replace the fictitious soil pressure, by dividing the factored overturning moment by one-half the footing width in the direction of the moment (see Figure 5-1).

5.3.5.4 Sliding

Although ASCE 7 does not require a check for foundation sliding, a common petrochemical industry practice is to use a factor of safety of 1.0, considering only 0.9 times the dead load for resistance. Sliding should be checked using the same load combinations that are used for overturning. The calculated overturning moment and shear that causes sliding should not be reduced by factors given in ASCE 7-16, Section 12.13.4, for the foundation sliding checks if IBC Alternative ASD Load Combinations are used. Lateral loads on spread footings and mats are resisted by friction between the footing and soil and passive pressure on the footing. A significant amount of lateral deflection is usually needed to mobilize full passive pressure of the soil. If the structure or its supported element cannot withstand the amount of lateral deflection required to mobilize full passive

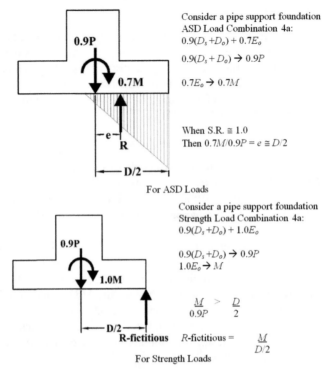

Figure 5-1. Concrete foundation design when stability ratio (S.R.) approaches 1.0.

pressure, then a partial passive pressure should be considered. Side friction (friction on the vertical side of the foundation) is not dependent upon the lateral movement; however, it is dependent upon a reliable contact surface. Soils that are subject to shrinkage owing to moisture content fluctuations may lose contact with the foundation. In this case side friction may not be relied upon for lateral resistance.

5.3.5.5 Pile Foundations

Pile foundations should have the capacity to resist the seismic load effect due to inertia forces from the superstructure and the foundation. The foundation weight includes the weight of the pile caps, grade beams, and soil weight directly above the pile cap and grade beams. When evaluating seismic load effect at the bottom of the pile cap (i.e., the pile head) the foundation weight should be included. It should not be considered when evaluating seismic load effect at the top of the pile cap (i.e., the column bases).

The foundation should be treated as a separate element from a flexible superstructure when combining seismic loads because they do not necessarily respond in phase with each other. The inertial force of the foundation should be defined as that imposed on a rigid element in the nonbuilding structures chapter of ASCE 7. The total seismic shear force on the piles and grade beams for a nonrigid structure can be determined by combining the base shear force of the superstructure with the foundation inertia forces by the square root of the sum of the squares (SRSS) as show in Equation (5-1).

$$V_{total} = \text{total lateral seismic load on piles} = \left(V_{fdn}^2 + V_{struct}^2 \right)^{0.5} \quad (5\text{-}1)$$

where $V_{struct} = V = $ Base shear of the superstructure

$V_{fdn} = 0.3S_{DS}\, I_e\, W_{fdn} = $ Seismic shear due to the foundation

$W_{fdn} = $ Weight of the pile cap, grade beams, and soil above the pile cap and grade beams

The seismic shear due to the foundation should not affect the vertical distribution of the seismic shear due to a flexible superstructure because the foundation is relatively stiff and usually responds out of phase with the super-structure. The flexible superstructure base shear and vertical distribution of this base shear should be calculated as if it were anchored directly to foundation soil at the base of the columns. If significant, the couple forces from the foundation seismic shear acting at the center of gravity of the foundation weight should be added by SRSS to the flexible superstructure couple forces.

In some cases, however, where the superstructure and foundation are both relatively stiff, such as a turbine or compressor on a heavy block foundation, the seismic force on the piles should be the algebraic sum of the component forces as shown in Equation (5-2) because the foundation will respond in phase with a rigid superstructure or equipment.

$$V_{total} = 0.3S_{DS}I_e\left(W_{equip} + W_{fdn}\right) \quad (5\text{-}2)$$

where $W_{equip} = $ Weight of rigid equipment or superstructure

Lateral loads on pile foundations are resisted by passive pressure on the piles and pile cap and side friction on the pile cap. The lateral deflection of the pile cap that is required to mobilize passive pressure on the pile cap should be compatible with the piles' strength capacity at that deflection. If the piles do not have enough strength to handle the deflection required to mobilize full passive pressure on the pile cap then partial passive pressure on the pile cap should be considered.

The lateral load capacity of a pile or pile group is essentially based on permissible horizontal deflections. Geotechnical consultants often set lateral load capacity based on a pile deflection at ground level varying from 1/4 to 1 in. The structural engineer should determine from the geotechnical consultant whether lateral pile load capacity is based on deflection or soil strength. If deflection is the governing criterion, the engineer should then evaluate whether a larger horizontal deflection is acceptable for the particular facility being designed, to avoid unnecessary conservatism.

Most petrochemical facilities' operating units are paved with 4 to 6 in. (10 to 15 cm) of concrete, reinforced with welded wire fabric to minimize shrinkage cracking. Consideration may be given to taking credit for the presence of the concrete paving in resisting lateral seismic loads to reduce the number of piles required for individual foundations only when the paving can be shown to be an integral, permanent, and structurally sound component of the overall foundation. While this may increase the paving thickness to create the necessary mass, this may be a cost-effective solution. Care should be exercised when combining lateral pile resistance, passive pressure on the pile cap, and frictional resistance of paving. Taking the maximum value of each mode of resistance may not be appropriate when used as a combination. This subject should be discussed with a geotechnical consultant. Caution should be taken to consider the effect of isolation joints when considering paving resisting seismic shear. The amount of lateral movement that the joint can accommodate should be considered in calculating resistance. Asphalt paving should not be considered as performing in the same manner as concrete paving. Asphalt paving is unreinforced and has low shear strength and therefore cannot contribute to resistance.

Pile Foundation Recommendations for Seismic Design Category C

Individual pile caps, drilled piers, or caissons should be interconnected by horizontal tie beams that are capable of resisting a minimum axial force equal to 10% of S_{DS} times the larger pile cap or column-factored dead plus factored live load. Tie beams are not necessary where equivalent restraint can demonstrably be provided by reinforced concrete beams within slabs on grade or reinforced concrete slabs on grade, or confinement by competent rock, hard cohesive soils, or very dense granular soils. In petrochemical plants, minimizing the use of tie beams is desirable to avoid interferences with underground piping systems and underground electrical duct systems. Adequate lateral restraint can, in many cases, be provided by passive soil resistance against the sides of the piles and pile caps if the piles can be shown to resist a lateral loading equal to 10% of S_{DS} times the larger of the pile cap or column factored dead and live load in addition to the design loads.

All concrete piles and concrete-filled piles should be connected to the pile cap by embedding the pile reinforcement in the pile cap for a distance equal to the full development length for compression or tension. No reduction of length for excess area should occur in the case of uplift. Anchorage of piles to the pile cap should be made by mechanical means other than concrete bond to bare steel for piles that resist uplift forces.

A minimum of four longitudinal bars, with a minimum longitudinal reinforcement ratio of 0.0025, should be provided over a minimum reinforced pile length plus the tension development length for augered concrete piles, metal-cased concrete piles, or uncased cast-in-place drilled piers. The minimum reinforcement length should be taken as the greater of (a) one-third of the pile length, (b) a distance of 10ft (3m), (c) three times the pile diameter, or (d) the flexural length of the pile. The flexural length of the pile should be taken as the length from the bottom of the pile cap to a point where the concrete section cracking moment multiplied by a resistance factor 0.4 exceeds the required factored moment at that point.

Concrete-filled pipe piles should have a minimum reinforcement of 0.01 times the cross-sectional area of the pile concrete. This reinforcement should be provided in the top of the pile with a length equal to two times the required cap embedment anchorage into the pile cap. Reinforcement should be provided for the full length of the pile.

Closed ties or equivalent spirals of minimum 3/8 in. diameter should be provided to confine the transverse reinforcing for augered concrete piles, metal-cased concrete piles, concrete-filled pipe piles, or uncased cast-in-place piles. Within a distance of three pile diameters from the bottom of the pile cap, transverse reinforcement spacing should not exceed 8 longitudinal-bar diameters or 6 in. (150 mm). Transverse reinforcement spacing throughout the remainder of the reinforced length of the pile should not exceed 16 longitudinal-bar diameters. A minimum diameter of 3/8 in. (9 mm) should be provided for transverse reinforcement consisting of closed ties or equivalent spirals. Hoops, spirals, and ties should be terminated with seismic hooks as defined in ACI 318.

Reinforcement for the upper 20 ft (6 m) of precast, prestressed piles should have a minimum volumetric ratio of spiral reinforcement not less than 0.007 or the amount required by the following equation:

$$\rho_s = 0.12 f_c'/f_{yh} \qquad (5\text{-}3)$$

where

ρ_s = Volumetric ratio (volume of spiral/volume of core),

f_c' = Specified compressive strength of concrete, psi (MPa), and

f_{yh} = Specified yield strength of spiral reinforcement, which should not be taken as greater than 85,000 psi (586 MPa).

A minimum of one-half of the volumetric ratio of spiral reinforcement required by Equation (5-3) should be provided for the remaining length of the pile.

Pile Foundation Recommendations for Seismic Design Categories D, E, and F

Pile foundations for structures assigned to Seismic Design Categories D, E, and F should meet all the recommendations of Seismic Design Category C and the following recommendations.

Foundation forces resisted by a pile group consisting of both vertical and batter piles should be distributed in accordance to their relative horizontal and vertical rigidities and the geometric distribution of the piles within the group. Batter piles and their connections should be capable of resisting forces and moments from load combinations that include the overstrength factor. Limited information on battered piles shows that the possibility exists that battered pile groups fail and no longer can support gravity loads after a major earthquake. The use of battered piles to carry gravity load is discouraged because of this possibility. Vertical piles should be designed to support the gravity load alone, without the participation of the battered piles. The analysis of systems with battered piles should not be modeled as a simple truss system, neglecting the bending moment induced by the batter. ASCE 7 recommends that forces be distributed to the individual piles in accordance with their relative horizontal and vertical rigidities and the geomantic distribution of the piles within the group.

A minimum of four longitudinal bars with a minimum reinforcement ratio of 0.005 should be provided for a minimum reinforced length plus tension development length for uncased cast-in-place drilled or augered concrete piles and metal-cased concrete piles. The minimum reinforced length should be taken as the greater of (a) one-half of the pile length, (b) a distance of 10 ft (3m), (c) three times the pile diameter, or (d) the flexural length of the pile, which should be taken as the length from the bottom of the pile cap to a point where the concrete section cracking moment multiplied by a resistance factor of 0.4 exceeds the required factored moment at that point.

Transverse confinement reinforcement should be provided throughout the reinforced length of the pile in accordance with ACI 318, Section 18.7.5.2 through 18.7.5.4. This reinforcement should be provided for the minimum reinforced length for uncased cast-in-place drilled or augered concrete piles and metal-cased concrete piles. For precast concrete piles this reinforcement should be provided for the full length of the pile.

For piles located in Site Classes A through D, longitudinal and transverse confinement reinforcement, as described above, should extend a minimum of seven times the pile diameter above and below the interfaces of soft to medium stiff clay or liquefiable strata. Longitudinal and transverse confinement reinforcement, as described above, should extend the full length of the pile in Site Classes E or F.

Precast, prestressed pile reinforcement should be provided in accordance with ASCE 7-16, Section 14.2.3.2.6.

The design of pile anchorage into the pile cap should consider the combined effects of bending moments due to fixity of the pile cap and axial forces due to uplift. Anchorage into the pile cap should be capable of developing the following: (a) in the case of rotational restraint, the lesser of the full shear, axial and bending nominal strength of the pile, or the shear and axial forces and moments from the overstrength load combinations; (b) in the case of uplift, the lesser of the axial tension resulting from overstrength load combinations, or the nominal tensile strength of the pile, or 1.3 times the pile pullout resistance. The pile pullout resistance should be the ultimate adhesive or frictional force that can be developed between the pile and the soil plus the weight of the pile.

The nominal strength of a pile section should be developed at splices of pile segments. If the splice has been designed to resist shear, axial, and moments from load cases with overstrength factor then the splice does not need to develop the nominal strength of the pile segment.

Interaction of the pile and the soil should be considered when establishing pile design shears, moments, and lateral deflections. The pile may be assumed to be flexurally rigid with respect to the soil if the ratio of depth of embedment to pile diameter or width is less than or equal to 6.

Vertical nominal pile strength should include pile group effects where pile center-to-center spacing is less than 3 pile widths or diameters. Where center-to-center spacing is less than 8 pile diameters, pile group effects should be considered on lateral pile nominal strength.

Steel piles should be seismically compact in accordance with AISC 341. Uplift connectors for steel piles should be mechanical connectors located within the pile embedded area. The use of mechanical attachments and welds over a length of pile below the pile cap equal to the depth of cross section of the pile should be avoided because a plastic hinge is expected to form in the pile just under the pile cap or foundation.

5.3.5.6 Pier and Tee Support Foundations

Pier and tee support foundations are typically used to support grade-mounted horizontal vessels, heat exchangers, miscellaneous piping, and cable trays. These foundations are treated as "inverted pendulums." As an inverted pendulum, the supporting column of the pier or tee support must meet the requirements of ASCE 7-16, Section 12.2.5.3, which requires the inverted pendulum to be designed for the bending moment at the base with the moment varying uniformly to a moment at the top equal to one-half of the calculated bending moment at the base. Please note that the vessel, heat exchanger, piping, or cable tray attached or supported at the top of the pier or tee support foundation and any connection between the supported item and the pier or tee support foundation are not designed for this moment.

The vessel or exchanger is anchored at one pier and allowed to slide at the opposite pier to relieve thermal expansion forces. The anchor pier is designed to resist all longitudinal seismic forces. For simplicity, both the anchor pier and sliding pier are designed the same. When this approach creates excessively large pier foundations, two alternatives should be considered. For vessels with low design temperatures, less than about 150 °F (65 °C), the vessel can be anchored at both ends. Here each pier would be designed for one-half the vessel expansion and one-half the seismic load. When thermal expansion cannot be accommodated without allowing the vessel to slide, the piers can be tied together with structural steel struts, so that again each pier foundation resists one-half the seismic load.

A study on longitudinal seismic loads on some analytical models (Richter unpublished) indicates that a significant amount of energy is absorbed at the sliding end pier, depending on the friction coefficient. This effectively reduces the load applied to the fixed end pier. In the study, a series of calculations were performed to investigate the seismic force distribution to pier foundations including nonlinear time history analyses on a two-dimensional model of a typical pier foundation. Table 5-5 gives tentative recommendations for load distribution to the fixed (V_{FIX}) and sliding ends (V_{SLIDE}), as a fraction of the total horizontal seismic load (V_{TOT}). Additional savings may be achieved by

Table 5-5. Recommendations for Load Distribution.

Friction Coefficient at Sliding End	V_{FIX}/V_{TOT}	V_{SLIDE}/V_{TOT}
0.0–0.2	1.0	0.6
>0.2	0.7	0.6

designing each pier separately, for the horizontal load distribution shown in the table.

Transverse seismic loads are usually carried by both saddles, and the loads are transmitted in shear to the anchor bolts. When determining the exact mass distribution at each saddle is not possible, some engineers apply 60% of the transverse seismic load to each saddle.

5.3.5.7 Vertical Vessel Foundations

The buckling criteria used for vertical vessels and their skirts affect the response modification factor, *R*, used for determining seismic loads on vertical vessel foundations (see Appendix 4.C). The engineer responsible for the foundation design is recommended to become familiar with and provide input to the seismic design criteria used by the engineer responsible for the vessel and skirt design prior to purchasing the vessel. This is especially important for long lead time vessels, heavy vessels, and revamp vessels that are reusing an existing foundation.

Historically, the foundation anchor bolts for tall vertical vessels and stacks have tended to stretch beyond yield when subjected to strong ground motion. Yielding of anchor bolts probably prevented collapse of these vessels. Based on this experience, these anchor bolts should be designed with ductile embedment into the foundation (see Section 5.3.5.2). Special care should be taken to avoid oversizing the anchor bolts. Excessively oversized anchor bolts could remain elastic during a seismic event, creating overturning moments in the foundation beyond that used in the design. This also leads to other detrimental behavior such as buckling of the skirt.

5.3.5.8 Earth Retaining Structures

Earth retaining structures are designed for a wide variety of applications in petrochemical facilities. Among some common uses are

- Free-standing retaining walls for vertical grade offsets;
- Walls of underground pits, sumps, and vaults; and
- Basement walls for buildings and similar structures.

General requirements for seismic design of earth retaining structures are provided in Section 15.6.1 of ASCE 7-16. Lateral earth pressures for Seismic Design Categories D, E, and F must be developed through a geotechnical analysis prepared by a registered design professional. This requirement recognizes the potential significance of dynamic soil pressure at sites with high ground motions and occupancy criteria, with dependencies on local soil parameters, potential geological hazards, groundwater, and the physical features of the retaining structure.

ASCE 7 allows earth retaining structures to be considered either yielding (unrestrained wall movement) or nonyielding (restrained wall movement) when subjected to lateral seismic design loads. A lateral displacement of 0.002 times the wall height is typically adequate to develop the minimum active pressure state. Generally free-standing or cantilever retaining walls can be considered as yielding walls. Nonyielding walls mobilize more soil pressure due to greater interaction between the wall and soil during seismic excitation.

Published methods and formulae are available for calculating dynamic soil pressure. A common procedure for the design of gravity retaining walls is that proposed by Mononobe and Okabe (M-O), referenced in Seed and Whitman (1970). Mononobe and Matsuo (1929) and Okabe (1926) modify Coulomb's classical solution to account for inertia forces corresponding to horizontal and vertical accelerations acting at all points of an assumed wedge failure. Backfill thrust against a wall is expressed as a function of the unit weight and slope of the backfill, the height of the wall, and the active stress coefficient. The active stress coefficient is a function of the friction angle of the backfill soil and the friction angle between the backfill and the wall. Seed and Whitman (1970) provide a simplified version of the original equation, which is commonly used in practice for horizontal backfill. The M-O equation is based on several assumptions, including the backfill being able to deform sufficiently to mobilize full shear resistance along the failure plane in the active zone and the accelerations are constant throughout the failing wedge.

A useful discussion and presentation of the Seed and Whitman (1970) formulae for yielding walls and the Wood (1973) method for nonyielding walls are presented in Section 7.5.1 of FEMA 450, Part 2: Commentary (FEMA 2003). The peak ground acceleration in this reference is taken as $S_{DS}/2.5$. ASCE 7-16 provides values for peak ground acceleration without the need to approximate the value.

SSI computational methods consider the dynamic properties of the soil, nonlinear soil behavior, the specific characteristics of the ground motion, and the inertial effects of the wall and connected superstructure. SSI is considered capable of producing more accurate results than limiting equilibrium design methods. Published studies are available that correlate SSI results to the traditional M-O derivations. Ostadan (2005) presents a method for building walls (nonyielding backfill), using a single degree of freedom system to model a rigid wall and foundation system. It considers the dynamic backfill properties, design motion, amplification of the motion in the backfill, and soil nonlinearity in developing the seismic soil pressure. Results from this method were compared extensively to more detailed SSI computations and shown to be accurate, although slightly conservative.

Chapter 19 of ASCE 7 contains optional, simplified criteria that allow for SSI effects in cases where the mathematical model for computing the seismic loads utilizes fixed base conditions. By applying these criteria, the base shear and distributed seismic loads can be reduced up to 30%. This chapter includes SSI provisions for both the equivalent lateral force procedure and the modal analysis procedure.

The response modification factor (R) for use in the design of earth retaining structures is dependent on the type and function of the structure. Earth retaining walls that are integral with buildings or structures and part of the seismic force-resisting system should be designed using seismic loads developed with the appropriate R value for the structural system utilized. Lateral soil pressure, both static and the dynamic increment from the seismic event, should be applied in loading combinations in accordance with code requirements for H in Chapter 2 of ASCE 7-16. Free-standing walls should be designed as nonbuilding structures not similar to buildings in accordance with Section 15.6.1 of ASCE 7-16. An R value of 1.25 and other seismic coefficients is acceptable for calculating the inertial loads considering the retaining wall to be in the category of "all other self-supporting structures" in Table 15.4-2.

5.4 STRUCTURAL DETAILS

5.4.1 Introduction

Section 5.3 discusses typical structures found in petrochemical plants. In general, specific seismic details used in the building industry also apply to petrochemical plants. Most of the building industry practice can be found in ASCE 7. This section discusses special seismic detailing requirements unique to petrochemical plants.

5.4.2 Steel

5.4.2.1 General

ASCE 7, AISC 360, and AISC 341 present the general seismic requirements for steel structures. Special details that are not commonly encountered in the building industry will be discussed here.

No special structural steel seismic detailing requirements for petrochemical structures are specified in Seismic Design Category A. Structural steel petrochemical structures in Seismic Design Categories B or C do not require special seismic detailing if a response modification coefficient, R, of 3 or smaller is used. Seismic detailing of structural steel in accordance with the AISC 341 is required for structural steel petrochemical structures in Seismic Design Category D or higher, unless a much lower R value (1.5 for braced frames or 1.0 for moment frames) is used in accordance with ASCE 7-16, Table 15.4-1.

The steel used in lateral force–resisting systems should be limited to those listed in AISC 341. See Section 5.2.3.2 in this document for commonly specified materials for structural steel and testing requirements. In extremely cold regions, fine-grained steels should be considered to reduce the likelihood of brittle failure under dynamic loads. ASTM A709 (ASTM 2018), *Standard Specification for Structural Steel for Bridges,* specifies impact test requirements for service temperatures as low as −60 °F (−51 °C). The bridge steel grades in ASTM A709 are cross-referenced to those commonly specified by AISC 360.

ASCE 7 includes many design provisions that affect structural steel design, such as seismic provisions for collector elements and diaphragms. For instance, collector elements, splices, and their connections to resisting elements should be designed for load combinations that include the overstrength factor, Ω_o. Collector elements are elements that transfer seismic forces to the lateral force–resisting system from other parts of the structure.

Structural steel lateral force–resisting systems consist primarily of moment frames and braced frames. The next two sections provide recommendations for seismic design of these structural systems.

5.4.2.2 Recommendations for Moment Connections

The testing completed by the SAC Joint Venture found that improved performance of pre-Northridge Earthquake (1994) connections into the inelastic range can be obtained by proper detailing of the beam flange welds, the use of continuity plates, the use of notch-tough weld metal, and proper detailing of the weld access hole. A discussion of detailing moment connections to resist seismic loads is provided in FEMA 350 (FEMA 2000a) and FEMA 353 (FEMA 2000b). As the AISC 341 Commentary notes, published testing, such as that conducted as part of the SAC project and reported in FEMA 350 and 355, may be used to satisfy the AISC conformance demonstration requirements subject to the approval of the Authority Having Jurisdiction (see AISC 341 Commentary E3.6c). A prescriptive procedure for designing prequalified reduced beam section and bolted end-plate moment connections can be found in AISC 358 (AISC 2016). Further guidance for the design of bolted end-plate moment connections and the required column stiffeners can be found in AISC Design Guides 4 (AISC 2003) and 13 (AISC 1999), respectively.

The two types of connections recommended for use in areas of high seismic risk are the bolted, end-plate moment connection (Figure 5-2) and the reduced beam section (RBS) moment connection (Figure 5-3). Tests have shown that these two connections provide the required ductility when subjected to earthquake type loading. Detailing requirements for these connections can be found in AISC 341. ASCE 7 requires AISC 341 to be followed for steel moment frame structures in Seismic Design Category D or higher unless an $R=1.0$ is used in the seismic load calculations in accordance with ASCE 7-16, Table 15.4-1. The provision of ASCE 7-16, Table 15.4-1, Note h, adds additional height exceptions for piperacks utilizing bolted end plate moment connection.

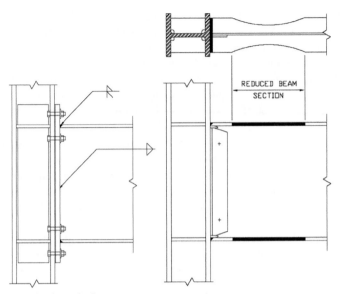

Figure 5-2. Bolted, end-plate moment connection.

The decision to use a lower R value for moment frames, thereby avoiding seismic detailing, may be economical in regions of moderate seismic risk. Seismic detailing makes connections and some members larger, which results in higher engineering costs, fabrication costs, and construction costs. One disadvantage of using the lower R value is that the foundations are designed for larger seismic loads. In regions of high seismic risk, designing the moment frames with a lower R value may not be economical because the resulting seismic loads may be so large that they outweigh the benefits of avoiding seismic detailing.

Some considerations should be made when planning to detail a special moment frame for seismic provisions. First, do not oversize the beam. A larger beam will result in a larger moment connection and columns when using the strong column weak beam concept. During the layout stage, make provisions for the lateral support of flanges of columns and beams when required. This may require adding additional struts or bracing. Consider making only selected beams within a moment frame to have moment connections. Beams that must be deeper to support large loads from equipment are good candidates for being designed as pin connected to the columns to avoid large moment connections. Be aware that vertical stiffeners required for stiffened end-plate moment connections may encroach into pipeways or may become a tripping hazard on platforms. RBS moment connections include many design considerations that are difficult to meet for petrochemical structures. For instance, the reduced beam section should have no connections, attachments, or welded items. Ordinary moment frames of steel designed to the seismic provisions are not subject to the foregoing limitations but much lower R values must be used for their design.

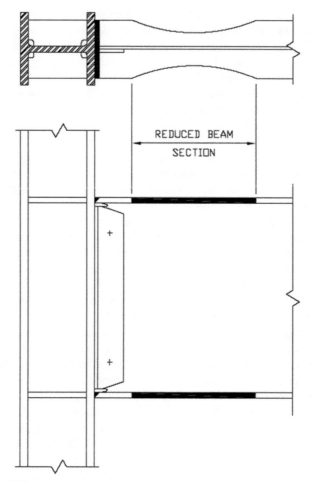

REDUCED BEAM
SECTION

Figure 5-3. RBS moment connection.

5.4.2.3 Recommendations for Braced Frames

In the petrochemical industry, braced frames rely on diagonal members to resist horizontal forces and to provide stability. These extra members tend to restrict access. Therefore, keeping the number of braces to a minimum and properly locating those that are used is good practice. In general, chevron (inverted V-type bracing) or diagonal bracing is used in preference to X-bracing in areas where seismic detailing is not required. In areas where seismic detailing is required, strong consideration should be given to X-braces because of the additional onerous detailing requirements for chevron bracing.

X-braces perform better than chevron or single diagonal braces when subjected to seismic loads. Single diagonal braces have less redundancy than X-braces. If the compression brace of a chevron brace buckles, then the beam should resist the vertical component of the tension brace in flexure. This force can be very large. One alternative is to transfer the unbalanced force to adjacent (upper or lower) chevron braces with the addition of zipper columns. Space for zipper columns should be dedicated during the equipment and piping layout stage.

K-braces, knee braces, or any bracing configuration that delivers a large unbalanced force into a column should be avoided and are not allowed in areas of high seismic risk or for structures in Seismic Design Category (SDC) D or higher in accordance with AISC 341.

Braced frames that are designed in accordance with AISC 341 require that SCBF connections be designed for the tensile capacity of the brace. In these cases, it is best to keep the size of the brace to a minimum so that connections do not become too large. Where compression braces are considered, use shorter spans for the braces. Consider using tension-only braces for OCBFs.

ASCE 7 requires structural steel braced frames in Seismic Design Category D or higher to be designed in accordance with AISC 341 unless they are designed as an ordinary braced frame with a $R = 1.5$, used for calculating the seismic base shear in accordance with ASCE 7, Table 15.4-1. The decision to use a lower R value for braced frames or braced bays, thereby avoiding seismic detailing, may be economical in regions of moderate seismic risk. Seismic detailing makes connections and some members larger, which results in higher engineering, fabrication, and construction costs. One disadvantage of using the lower R value is that the foundations are designed for correspondingly larger seismic loads. In regions of high seismic risk, designing the braced bay or braced frames with a lower R value may not be economical because the resulting seismic loads may be so large that they outweigh the benefits of avoiding seismic detailing.

The location of vertical and horizontal bracing should be worked in with the equipment location and access requirements early on in a project so that a continuous and positive load path exists for seismic loads.Bracing should be located so as to avoid interference with piping, ducts, electrical trays and conduits, and accessways. Also, diagonal bracing should be arranged so that it will not distribute lateral load to only a few foundations.

If a braced frame is detailed in accordance with AISC 341, then the brace member size should be the minimum size to meet the requirements. This is important because if the brace size increases, then the brace connection also increases. Use braces that are of a smaller length to reduce the impact of slenderness requirements. Consider tension-only braces for ordinary concentric braced frames. During the layout stage, make provisions for lateral bracing requirements. Eccentrically braced frames have many lateral bracing requirements that make them impractical for petrochemical structures.

5.4.2.4 Pipeways

To facilitate access requirements, pipeways are generally designed as moment-resisting frames in the transverse direction and are braced in the longitudinal direction. Longitudinal bracing transfers seismic and other longitudinal forces to the foundations. Normally, a long pipeway would require expansion joints (slotted strut connections) to allow for thermal movements, and thus the longitudinal bracing is designed for the tributary segment between expansion joints [typically, 100 to 200 ft (30 to 60 m), but this depends on plant and site conditions]. Pipeway struts are considered collector elements and should be designed for load combinations that include the overstrength factor, Ω_o. Pipeway struts typically do not frame into the columns at the same elevation as the transverse beams. This will introduce some additional column design requirements for SMF because of the lack of lateral support at the connection.

5.4.2.5 Air Coolers (Finfans)

Typically, air coolers are supported on pipeways. The air cooler is generally composed of a large box containing the fans and motors, supported on four or more legs. This generally creates a condition where a relatively rigid mass is supported on flexible legs on top of a pipeway. The support legs should be braced in both directions. Knee braces should be avoided. Vertical bracing should intersect columns at panel points with beams. Chevron bracing can be used as long as headroom is not compromised. Whenever possible, the air cooler is recommended to be designed without vendor-supplied legs and be supported directly on the piperack structural steel.

Note that the structural steel directly supporting the air coolers should be designed to the same level of seismic detailing required of the pipeway structural steel. This should be the case regardless of whether the air-cooler vendor or the engineering contractor provides the supporting steel. In the case of the air-cooler vendor providing the supporting steel, the seismic detailing requirements should be made clear during the request-for-quotes stage.

Though not a seismic design issue, care should be exercised to prevent possible resonance effects due to the operating frequency of the air-cooler motors being too close to the natural frequency of the support structure.

5.4.3 Concrete and Foundations

5.4.3.1 Framed Concrete Structures

The connections of ductile moment-resisting frame concrete structures, when detailed and constructed properly, have been observed to suffer little or no failure during a large seismic event. In petrochemical facilities, this includes structures such as table-top structures and concrete structures supporting equipment or other steel structures.

However, clearly some of these items should be designed so that the connected elements can perform properly. This requires that the joints be proportioned and detailed to allow the elements framing into them to develop and maintain their strength and stiffness while undergoing large inelastic deformation.

5.4.3.2 Equipment Foundations

In general, most equipment foundations are designed with the equipment considered as rigid. Therefore, for seismic conditions, the anchorage should be designed for the code-specified shear. Additional ties are required around the anchorage in the top of columns or pedestals in accordance with ASCE 7. Closely spaced ties are recommended to be placed at the base of pedestals that transfer seismic loads to the foundation through flexure like a cantilever column. Special consideration should be given to vessels and exchangers as outlined below.

5.4.3.3 Horizontal Vessels and Exchangers

Horizontal vessels and exchangers are normally supported on two pedestals supporting saddles that conform to the vessel curvature. One saddle is fixed to the pedestal, and the other is allowed to slide for thermal expansion. Figure 5-4 shows special tie details for foundations in high seismic design categories. Additional consideration should be given to multiple tie sets to provide adequate clearance for aggregate to pass between the ties. For additional guidance, refer also to Section 5.3.5.6.

5.4.3.4 Vertical Vessels and Stacks

Skirt-supported vessels are sensitive to buckling failures. ASCE 7-16, Section 15.7.10.5, contains provisions to ensure that skirt-supported vessels do not fail prematurely from buckling. If the skirt-supported vessel is classified as a Risk Category IV structure or if an R value of 3 is used, ASCE 7 requires the structure to be designed for seismic loads based on I_e/R equal to 1.0. This requirement ensures that the vessel and skirt support remain elastic during the design level earthquake. While the vessel and skirt support are allowed to be evaluated at critical buckling or yield as appropriate (factor of safety equals 1.0) for the $I_e/R = 1.0$ load case, this

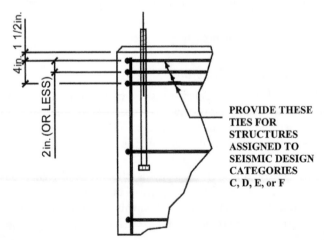

Figure 5-4. Special tie details for anchorage.

load case will govern the design of the support skirt in most cases and will occasionally govern the design of the vessel.

ASCE 7-16, Table 15.4-2, contains multiple entries for skirt-supported vessels that have proven confusing to users. Typical skirt-supported vertical vessels fall under the type "All steel and reinforced concrete distributed mass cantilever structures not otherwise covered herein, including stacks, chimneys, silos, skirt-supported vertical vessels; single-pedestal or skirt-supported." The value of R for a skirt-supported vessel will be either 2 or 3 depending on whether or not the special detailing requirements are met. To justify an R value of 3, special detailing requirements must be met because skirt-supported vessels are sensitive to buckling failures as described above.

A review of ASCE 7-16, Table 15.4-2, shows that an R value of 3.0 may be used if the skirt-supported vessel is designed to the provisions of ASCE 7-16, Section 15.7.10.5, which requires elastic design of the skirt. If the provisions of ASCE 7-16, Section 15.7.10.5, are not followed (vessels in Risk Categories I, II, or III only), an R value of 2.0 must be used. In general, the increased R value will only have an impact on the design of the vessel foundation. Use of an R value of 3.0 should be considered to reduce the cost of the foundation or to justify the use of an existing foundation. The total cost of the vessel and foundation must be determined for these evaluations.

Figure 5-4 illustrates some special tie details for anchorage. Spacing shown is appropriate for single ties (one tie per set). Spacing between ties should be increased to accommodate ties with more than one tie per set.

5.4.3.5 Foundations on Piles

Pile design data are typically reported for two head conditions: fixed and free. While actual head conditions usually have little effect on the compressive capacity of a pile, for most soil types, lateral displacement is often significantly less for a given shear load when the fixed head condition is provided in the foundation design. Rotational restraint of the pile head can be obtained by developing properly detailed reinforcing with sufficient moment capacity into the pile cap. Cast-in-place drilled shafts and auger cast piles can have their rebar cages detailed to extend above the top of the pile elevation for embedment into the pile cap. Piles that are driven to refusal, such as precast concrete and steel pipe or H-shape piles, are usually purchased longer than the design length requires and, after driving to refusal, are cut off to the correct top of pile elevation. For concrete piles, the usual practice is to demolish the concrete to the cutoff elevation and leave the deformed reinforcing and prestressing strand intact for incorporation into the pile cap. Hooked or straight deformed reinforcing can be welded to the tops of steel pipe and H piles with a flare groove weld. Reinforcing that conforms to the ASTM A706 specification is recommended to facilitate welding. Simply embedding a rebar cage into a concrete filled pipe pile may not provide the necessary rotational restraint unless true composite action can be obtained. Sufficient concrete cover should be provided below and around the welded bars. Single pile foundations should be considered free headed unless restrained by fixed end grade beams that connect them to other restraining foundations. Shear transfer at the pile head is often achieved by embedding the pile into the pile cap, 6 in. minimum is commonly specified. If cap-reinforcing mats are detailed to be placed above the pile cutoff elevation, concrete shear capacity of any edge piles in the group should be checked if only embedment is used for shear transfer to the pile.

5.4.4 Masonry

Reinforced masonry structures are often an integral part of the landscape throughout a petrochemical plant. Masonry structures are limited to items such as retaining walls and barriers, firewall barriers between equipment such as electrical transformers, operator shelters, auxiliary buildings, and in some cases control buildings. Often, masonry is used for thermal linings in furnaces, heaters, and other high-temperature equipment. In an active seismic zone, connection details for the aforementioned items should comply with the IBC, ASCE 7, and related masonry codes from The Masonry Society (TMS). However, care should be taken to ensure thermal growth compatibility of any ties with that of the masonry.

5.4.5 Timber

In the petrochemical industry, most structures are built with materials other than timber, such as steel and concrete. However, when timber is used for the construction of buildings or supporting systems, connectors similar to commercially available seismic ties or with good seismic resistance are recommended. In a high seismic area, these special connectors will provide excellent uplift and lateral load resistance to the structure.

5.4.6 Base Isolation

5.4.6.1 Introduction

The principle of base isolation is to provide a discontinuity between two bodies in contact so that the motion of either body in the direction of the discontinuity is decoupled. In building and refinery structures, the discontinuity is typically positioned at the base of the structure (hence, the term "base"), between the structure and the ground, and is characterized by a low shear resistance (relative to the structure and ground). The low shear resistance effectively lengthens the period of response of the structure significantly, increasing rigid body displacement motion, but significantly reducing structural accelerations and hence inertial loading. Thus, in a seismic event, severe loads from horizontal shaking are mitigated at the expense of increased global displacements. The increased lateral displacements are a minor disadvantage of isolator units, which are usually accommodated by provision of flexible joints in piping, cables, and secondary steel, such as walkways, that are attached to the primary structures.

Base isolators may be manufactured in several forms and in various materials, ranging from thin sliding surfaces, rubber bearings, mixed rubber and lead

Figure 5-5. Base-isolated LNG tank, Greece.

bearings, fiber-reinforced elastomeric bearings, and friction pendulum devices. Base isolators are commonly used on bridges, new buildings, and other large civil engineering structures, but have had limited application in petrochemical facilities. A notable exception to this is liquefied natural gas (LNG) tanks, where they have been used before. At the Revithoussa LNG import terminal in Greece, a partially below-ground design was used, where tanks pits were excavated out of rock for two 65,000 m³ tanks. Friction pendulum base isolation was utilized to reduce seismic input loading on the tanks (Figure 5-5).

The design of a specific base-isolation unit type is governed by permissible levels of deflections of the isolated structure and the necessary stiffness that the isolator must have to provide adequate rigidity under service loads (e.g., thermal and live loads), wind loading, and minor seismic loads.

5.4.6.2 Potential Advantages of Base Isolation

Base isolation is an innovative means of seismic protection increasingly used for critical facilities (disaster command and control/response facilities, hospitals, landmark and historic building structures, bridges, and now industrial facilities). Structures are isolated from ground motion shaking through the use of various systems, such as lead rubber combination isolators, friction pendulum isolators, or others.

The principal advantages of introducing some type of seismic isolation, and flexible pipe connections in association with it, are three-fold:

• Base isolation may achieve the desired safety objectives at a lower cost. While conventional design (i.e., usually an increase in strength and stiffness) becomes more expensive as the design motions increase, the probable costs of isolation and flexible joints remain unaltered or increase much more slowly (Figure 5-6). A potential economic advantage therefore exists.

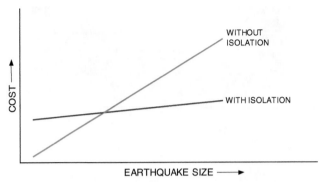

Figure 5-6. Relative cost of including base isolation in structural design.

- The second advantage is that conventional design only protects up to the arbitrary level adopted in the design, and a greater earthquake could always occur. Also, with time, society's perception of risk evolves, typically in the same direction; thus, authorities may ask in the future for increased levels of safety, leading to belated retrofits, which are often very costly and perhaps impossible. In this sense, incorporating some type of isolation in the design adds tolerance to beyond-design basis events, whether real or postulated. By their very nature, base-isolated concepts tend to establish upper bounds on the structural demands, somewhat independently of the event size. Hence, they add protection, not just for the events considered in the design basis, but also for the less probable events beyond it (Figure 5-7).

- A final advantage is that conventional design often has, as its performance basis, a criterion of "collapse prevention" rather than "continued operability," which is usually the performance criterion of isolated structures. The reason for this is usually one of economics. This is an important distinction that must be made when comparing conventional seismic design/strengthening with

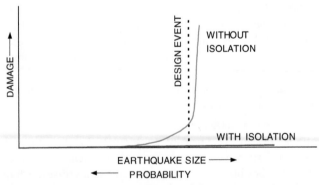

Figure 5-7. Effect of structural damage with earthquake size (with and without base isolation).

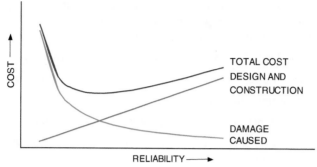

Figure 5-8. Variation in cost with reliability.

use of isolation. Namely, conventional earthquake engineering design often results in structures, which, while they may be designed not to collapse, may be irreparably damaged beyond repair during strong ground shaking. In contrast, isolated structures can readily be designed to function immediately following an earthquake. Hence, there is an increase in reliability for a given level of cost (Figure 5-8).

Against these advantages must be weighed isolator maintenance considerations, the increased deflections/drifts that the piping and other systems will need to accommodate, and the absolute requirement that the isolator remain functional over time, that is, it must "isolate" when an earthquake occurs. Also, base isolation does not isolate against vertical ground motions. Thus, while horizontal seismic forces are greatly decreased, vertical seismic forces are not.

5.5 PHYSICAL INTERACTION OF STRUCTURES AND COMPONENTS

5.5.1 Introduction

This section deals with the physical interaction between structures and components and the precautions to take to minimize the negative effects.

5.5.2 General

The physical interaction between structures and components of a petrochemical plant during a seismic event resembles that in the general commercial environment. The configuration of structures or components largely determines how seismic forces are distributed and also influences the relative magnitude of forces and displacements.

The objective of a good petrochemical facility configuration is high operating efficiency and low capital cost. Process, mechanical, and piping design requirements and the desire to minimize plot space encourage placing components close together. This inevitably creates the potential for damaging interaction of structures and components during an earthquake. The structural engineer must raise concerns about structure and component interaction.

Examples of potential structure and component interaction are

(a) Tall vertical vessels with platforms or pipe in close proximity to each other;

(b) Vertical or horizontal vessels with shared working platforms;

(c) Interconnecting pipeways;

(d) Mobile equipment such as davits, cranes, and so on; and

(e) Distributive systems that connect components supported on piles with components not supported on piles.

Some of the general concepts that are practiced to avoid or minimize physical interaction are as follows:

(a) Lay out all structures as regular structures whenever possible.

(b) Provide adequate space between components.

(c) Check the maximum displacement. Make sure there is enough room between the structure and component to match the displacement calculated.

(d) Provide an appropriate support and restraint configuration for pipes to minimize transfer of load across flanges and couplings.

(e) Make piping systems more flexible to mitigate the coupling effects of large pipes. Include loops in the pipes wherever needed.

5.5.3 Structure and Component Interaction

General practice is to lay out all structures and components to minimize any physical interaction during a seismic event. When platforms or vessels are in close proximity to each other, the displacements and/or periods should be checked to ensure that the spacing between such structures exceeds the sum of the absolute values of the displacements. The displacements (determined with code-mandated strength level forces) for this evaluation should be increased by C_d/I_e to account for inelastic behavior. Also refer to Section 5.3.3.2 herein, for further information on dealing with inelastic behavior. For nonstructural components such as piping, ASCE 7-16 also requires the design displacement be increased by the I_p factor.

Often, piping in structures is routed vertically from a higher elevation to a lower elevation into flanges and couplings at the nozzles of turbines, pumps, vessels, etc. Large displacements of these pipes may cause leaks at the interface, thus causing fires and in some cases leading to explosions. All effort should be made to provide adequate system flexibility in the pipes to minimize load transfer across flanges and couplings due to seismic loads. This may be achieved by several methods, the most practical being proper support of the piping, installation of properly sized snubber devices, and limits set on the displacement of pipes in the structure. Where piping is routed out of the upper level of a structure to rotating equipment or vessels located at grade, relative drift of the structure may impart unacceptable nozzle loads on the grade-mounted equipment and should be accounted for by increasing the flexibility in the piping system.

Section 15.7.4 of ASCE 7-16 requires, unless otherwise calculated, the minimum displacements in Table 15.7-1 to be assumed for design of piping systems connected to tanks and vessels. These movements are to be multiplied by C_d. For unanchored tanks and vessels, these minimum displacements are quite large and when located above the support or foundation elevation must be increased to account for drift of the tank or vessel and foundation movements due to anticipated settlement and/or relative seismic displacements. The seismic displacements in the piping system cannot significantly affect the mechanical loading on the equipment attachment nozzles; piping accessories that increase flexibility and help the system tolerate movement are allowed if properly designed for the seismic motion and operating conditions. For nonstructural components such as piping, ASCE 7-16 also requires the design displacement be increased by the I_p factor. Examples are expansion joints and bellows.

5.5.4 Pipeways

Often, pipeways intersect each other, thus care should be taken to avoid transferring longitudinal seismic displacement and forces from one pipeway into transverse loads on another. Adequate longitudinal bracing or physical separation of pipeways can protect against such effects. Layout of piping and cable tray systems should be done to minimize such effects. Pipe anchors designed for thermal restraint must also be designed to resist the longitudinal seismic force due to the weight of the piping system, because frictional resistance at those locations where the piping system is simply resting on supports is often relieved during an earthquake.

5.6 GEOTECHNICAL CONSIDERATIONS

Foundations for structures designed with the aid of modern building codes and seismic analysis techniques have generally performed well during major earthquake events. This satisfactory state of events is due to a combination of a good understanding of seismic loads, good construction and design procedures, and adequate safety factors used for static loads. Seed et al. (1991) and Krinitzsky et al. (1993) provide some excellent background materials on seismic-related foundation issues. The following sections provide additional discussion on selected geotechnical issues.

5.6.1 Piled Foundations

The ASCE Geo-Institute published several papers reviewing the present understanding of dynamic pile response (Prakash 1992). Among those papers was one by Hadjian et al. (1992) that reviewed the state of the practice as it relates to both code provisions and engineering practice. The following is largely abstracted from that paper and from Krinitzsky et al. (1993).

Little effort is typically expended on pile-soil-structure interaction unless difficult soil conditions exist, or the capital expense or importance of the structure is great. However, many procedures have been proposed and are in use for independent piles, pile groups, and linear and nonlinear soil responses (e.g., Prakash 1992; Gazetas 1991; Nogami et al. 1991, 1992). Considering pile-soil-structure interaction could potentially reduce the seismic demands and hence the required structural capacity. This is due to the incoherent motions between soil and the more rigid structure, the radial damping through foundation, and the reduced ground motions with depth.

Piles inserted into sloping grounds, or adjacent to slumping materials, may experience lateral loading due to the transverse movement of soil, in addition to the earthquake inertia load, during ground failure. Such conditions are difficult to design against.

Friction piles placed in loose sands, sensitive clays, or high-water-content clays may experience loss of capacity or settlement and dragdown forces during significant seismic events.

Battered piles offer a stiffer configuration to lateral loads than vertically oriented piles and therefore tend to increase seismic loads. Structures not appropriately designed or constructed for this response can experience severe damage. There are proposals to eliminate future construction with battered piles in seismic areas because of poor performance.

The seismic performance of piles can be improved by extending them to deeper, more competent soils, adding belled bases to improve tension and end-bearing capacity, or by adding a surface surcharge to increase the confining stress and strength at depth. Driving full-displacement piles densifies the surrounding soil, thereby improving performance.

5.6.2 Soil Strength Considerations

Several codes allow the use of increased soil strengths under the action of seismic ground shaking. The increase is typically on the order of one-third or greater. Some soil types, such as sensitive clay and loose saturated sand, may experience significant strength reduction under repeated cyclic loadings. This reduction should be considered in evaluating the foundation responses under earthquake ground shaking.

The larger factors of safety used in foundation engineering for static loads result in significant excess capacity that may be considered for the less frequently occurring transient loads due to seismic ground shaking. The general satisfactory performance of buildings and foundations designed under modern seismic codes would seem to validate this procedure. However, significant care must be exercised to ensure that the foundation response under dynamic cyclic loading has been considered in an analysis attempting to characterize it with simple static load methods typically used in designs.

5.6.3 Geotechnical Site Investigations

Site investigations are required to acquire basic foundation design data. Zones of potential seismic activity require some additional considerations that are necessary to quantify local site response characteristics and seismic hazards. The following is a partial list of data required for conventional foundation analyses and seismic response and hazard analyses:

(a) Depth to bedrock;

(b) Ground water table location;

(c) Soil stratification;

(d) Soil physical characteristics (grain size distribution, index properties, organic content, density, void ratio, water content, SPT blow counts);

(e) Soil mechanical characteristics (elastic properties, consolidation properties, shear strength);

(f) Soil profile classification, such as rock, stiff soil, and soft soil profiles;

(g) Seismic parameters and site amplification factors;

(h) Unique conditions (sloping bedrock, surface grade, and proximity to faults, cliff areas, river fronts or coastal areas, arctic conditions);

(i) Site investigation reports with descriptions of drilling, sampling, and in-situ testing methods, as this information in useful in aiding the comparison of results from different investigations;

(j) Lateral earth pressures on earth-retaining structures due to earthquake ground motions in SDC D, E, or F; and

(k) Assessment of the potential consequences of soil liquefaction and soil strength loss.

References

ACI (American Concrete Institute). 2014. *Building code requirements for structural concrete.* ACI 318. Farmington Hills, MI: ACI.

ACI. 2016a. *Building code requirements for masonry structures.* TMS 402. Farmington Hills, MI: ACI.

ACI. 2016b. *Specifications for masonry structures.* TMS 602. Farmington Hills, MI: ACI.

AISC. 1999. *Stiffening of wide-flange columns at moment connections: Wind and seismic applications: Steel design guide 13.* Chicago: AISC.

AISC. 2003. *Extended end-plate moment connections Seismic and wind applications: Steel design guide 4,* 2nd ed. Chicago: AISC.

AISC. 2016a. *Seismic provisions for structural steel buildings.* AISI 341. Chicago: AISC.

AISC. 2016b. *Specification for structural steel buildings.* AISC 360. Chicago: AISC.

AISC. 2016c. *Prequalified connections for special and intermediate steel moment frames for seismic applications.* AISC 358. Chicago: AISC.

AISI (American Iron and Steel Institute). 2016. *North American specification for the design of cold–formed steel structural members.* AISI S100. Washington, DC: AISC.

API (American Petroleum Institute). 2014. *Welded steel tanks for oil storage*, 12th ed. Washington, DC: API.

ASCE. 2002. *Specification for the design of cold–formed stainless steel structural members.* ASCE 8-02. Reston, VA: ASCE.

ASCE. 2010a. *Anchorage design for petrochemical facilities.* Reston, VA: ASCE.

ASCE. 2010b. *Structural applications for steel cables for buildings.* ASCE 19-10. Reston, VA: ASCE.

ASCE. 2014. *Design loads on structures during construction.* ASCE 37-14. Reston, VA: ASCE.

ASCE. 2016. *Minimum design loads and associated criteria for buildings and other structures.* ASCE/SEI 7-16. Reston, VA: ASCE.

ASTM. 2018. *Standard specification for structural steel for bridges.* ASTM 709. West Conshohocken, PA: ASTM.

AWC (American Wood Council). 2015. *Special design provisions for wind and seismic.* AWC SDPWS-15. Leesburg, VA: AWC.

AWC. 2018. *National design standard for wood construction.* AWC NDS-18. Leesburg, VA: AWC.

FEMA. 2000a. *Recommended seismic design criteria for new steel moment-frame buildings.* FEMA 350. Washington, DC: FEMA.

FEMA. 2000b. *Recommended specifications and quality assurance guidelines for new steel moment-frame construction for seismic applications.* FEMA 353. Washington, DC: FEMA.

FEMA. 2003. *National earthquake hazard reduction program recommended provisions for seismic regulations for new buildings and other structures—Part 1 Provisions and Part 2 Commentary.* FEMA 450. Washington, DC: Building Seismic Safety Council.

Gazetas, G. 1991. "Chapter 15—Foundation vibrations." In *Foundation engineering handbook*, H. Y. Fang, ed. New York: Van Nostrand Reinhold.

Hadjian, A. H., R. B. Fallgren, and M. R. Tufenkian. 1992. "Dynamic soil-pile-structure interaction: The state-of-the-practice." In *Piles under dynamic loading, geotechnical special publication no. 34*, S. Prakash, ed., 1–26. New York: ASCE.

IBC (International Building Code). 2018. *International building code.* Country Club Hills, IL: International Code Council.

Krinitzsky, E. L., J. P. Gould, and P. H. Edinger. 1993. *Fundamentals of earthquake resistant construction.* New York: Wiley.

Mononobe, M., and H. Matsuo. 1929. "On the determination of earth pressures during earthquakes." In *Proc., World Engineering Congress*, 9. Tokyo.

Nogami, T., H. W. Jones, and R. L. Mosher. 1991. "Seismic response analysis of pile-supported structure: Assessment of commonly used approximations." In *Proc., 2nd Int. Conf. on Recent Advantages in Geotechnical Earthquake Engineering and Soil Dynamics*, S. Prakash, ed. St. Louis.

Nogami, T., J. Otani, K. Konagai, and H. L. Chen. 1992. "Nonlinear soil-pile interaction model for dynamic lateral motion." *J. Geotech. Eng.* **118** (1): 89–106.

Okabe, S. 1926. "General theory of earth pressures." *J. Jpn. Soc. Civ. Eng.* **12** (1): 1277–1323.

Ostadan, F. 2005. "Seismic soil pressure for building walls—An updated approach." *J. Soil Dyn. Earthquake Eng.* **25** (7): 785–793.

PIP (Process Industry Practices). 2017. *Structural design criteria.* PIP STC01015. Austin, TX: PIP.

PIP. 2018. *Application of ASCE anchorage design for petrochemical facilities.* PIP STE05121. Austin, TX: PIP.

Prakash, S. 1992. *Piles under dynamic loading: Geotechnical special publication no. 34.* New York: ASCE.

Richter, P. J. Unpublished. *Nonlinear analysis to evaluate distribution of frictional forces to horizontal vessel support piers.* Irvine, CA: Fluor Daniel.

Seed, H. B., R. C. Chaney, and P. Sibel. 1991. "Chapter 16—Earthquake effects on soil-foundation systems." In *Foundation engineering handbook,* 2nd ed., H. Y. Fang, ed. New York: Van Nostrand Reinhold.

Seed, H. B., and R. V. Whitman. 1970. "Design of earth retaining structures for dynamic loads." In *Proc., ASCE Specialty Conf. on Lateral Stresses in the Ground and Design of Earth Retaining Structures,* 103–147. Ithaca, NY.

SJI (Steel Joist Institute). 2015. *Standard specification for K-series, LH-series and DHL series open web steel joists and for joist girders.* SJI-100. Florence, SC: SJI.

Wood, J. H. 1973. *Earthquake-induced soil pressures on structures.* Rep. No. EERL 73-05. Pasadena, CA: California Institute of Technology.

CHAPTER 6

Walkdown Evaluations
of Existing Facilities

6.1 INTRODUCTION

A "walkthrough" or "walkdown" evaluation is the term generally applied to an on-site, mostly visual, screening review where as-installed components are methodically "walked down" and evaluated for potential seismic vulnerabilities. Using this method, rapidly and cost-effectively identifying the highest risk areas and prioritizing further, more detailed evaluations that might be appropriate are possible. This chapter is intended to give practical guidance to engineers who will perform such evaluations.

6.2 BASIS FOR PERFORMING WALKDOWNS

Petrochemical and other industrial facilities must increasingly demonstrate safety against toxic releases and pollution by mandates such as the CalARP (2013) program requirements (see Section 4.6 and Appendix 4.F), the OSHA law (29 CFR 1910.119), and the EPA law (40 CFR Part 68). Currently, no widely used and accepted standard is available for evaluating existing facilities for seismic loads. While some highly regulated industries, such as nuclear power plants, require conformance to new codes each time design standards change, when owners of petrochemical and other industrial plants evaluate facilities, it is generally done voluntarily. Upgrade and acceptance decisions rely on cost-benefit considerations, with criteria set by the owners in agreement with local authorities as necessary.

Walkdown evaluations have also proven to be beneficial when conducted after the construction of new facilities is completed. A walkdown evaluation should be incorporated into a prestartup safety review for new facilities.

Walkdown techniques along with limited analytical evaluations have been used extensively in California petrochemical facilities rather than a nuclear-type methodology to satisfy CalARP and similar mandates since the early 1990s. This approach has been implemented because

241

- Very few existing facilities can demonstrate conformance with current seismic codes. Seismic design codes change constantly to incorporate lessons learned from past earthquakes and ongoing research and have usually become more restrictive and more conservative.

- The walkdown method takes advantage of the lessons learned from the earthquake performance of industrial facilities, namely that most installations perform well, even when not designed specifically for seismic loads. The occurrence of damage can usually be traced to known causes that could have been mitigated by the evaluations described in this section.

- The walkdown method is cost-effective. The intent is for all components of interest to be looked at during the walkdown, but more costly detailed analyses are eliminated for all but the highest-risk items. Obvious problems can be quickly identified and mitigated, as can those areas where low-cost modifications or maintenance can significantly improve the seismic integrity of the equipment/structure (e.g., missing hardware).

- The method is logical and defensible to regulators and owners. It considers actual plant conditions; relies on demonstrated performance in past earthquakes, backed up by performance observations and data; and incorporates the experience, judgment, and common sense of the engineers performing the review. Several California regulators have accepted and recommended this approach for CalARP seismic assessments.

- The method is appropriate for regions of high and low seismicity. The level of seismic hazard can be accounted for in the walkdowns.

In summary, the walkdown methodology allows the owner to identify high-risk items and assess potential safety, pollution, and economic exposure due to seismic events.

6.3 GENERAL METHODOLOGY

Walkdowns are generally performed by an engineer or team of engineers in a methodical, systematic manner, to ensure consistency and completeness. The overall methodology may include several or all of the following elements:

- Meetings with owners, operators, regulators, process safety engineers, or other appropriate parties to discuss evaluation objectives and to establish the facility's performance requirements.

- Identification of equipment, structures, and piping of interest. If the review is being performed as part of a process hazards analysis or process-related safety review, the walkdown engineers should review the assumptions used in the hazard analysis regarding the expected post-earthquake availability of critical systems. If the review is voluntary, for purposes such as evaluating insurance needs or overall risk quantification, the engineer may be asked to review all major equipment and structures in a facility.

- Establishment of damage categories. Walkdown efforts are often required to be coordinated with a process safety team to establish requirements for use in the consequence analyses. For example, the process safety team may need an indication of whether, in a given earthquake, a vessel might
Be undamaged,
  ○ Suffer minor damage and leak, or
  ○ Fail catastrophically.
- Collection of site data. Data of interest include seismic hazard data, fault locations, available soil borings, plot plans to locate items, and seismic design basis data applicable to the unit being evaluated.
- Walkdown evaluation of components. This is done systematically, using checklists for each component, to document the evaluation and to serve as a reminder of the screening criteria. Figure 6-1 shows an example evaluation sheet.
- Review of drawings as necessary. This may be done to check adequacy of reinforced concrete structures; to verify anchorage details; or to identify configurations that cannot be visually reviewed due to obstructions, fireproofing, insulation, etc.
- Identification of items for analytical review. These may include "worst-case" items or any items that appear to be seismically vulnerable. This is based on the potential for damage that would cause toxic release, pollution, or other unacceptable performance characteristics, such as damage that would cause significant business interruption.
- Documentation of "poor" or "questionable" items for owners or regulators. Sufficient explanation must be provided such that corrective actions, maintenance, further evaluation, etc., will in fact address the engineer's concerns. The engineer must recognize that others may perform additional risk mitigation without further consultation with the engineer who performed the review.
- Recommendation of structural or mechanical fixes or other efforts that would mitigate risk from items listed above. The engineer may be required to interact with process safety engineers and owners to evaluate the economic and technical feasibility of structural and process modifications.
- Identification of consequences due to failure of each of these items and the prioritization of any recommended risk mitigation actions according to the consequences of failure and the existing level of risk.
A rating system may be appropriate, identifying
  ○ Major seismic vulnerabilities that should be fixed immediately;
  ○ Serious vulnerabilities that might require fixes, depending on economics; and
  ○ Relatively simple fixes that might be performed in conjunction with routine maintenance or during the next scheduled turnaround.

FIELD DATA SHEET FOR EQUIPMENT

EQUIPMENT ID:
DESCRIPTION:
LOCATION:

SCREENING EVALUATION: SUMMARY

Summary of Evaluation: ____Adequate ____Not Adequate
                                           ____Further Evaluation Required

Recommendations:

SCREENING EVALUATION: ANCHORAGE

Noted Anchorage Concerns:

____Installation Adequacy   ____Weld Quality
____Missing or Loose Bolts   ____Corrosion
____Concrete Quality   ____Other Concerns
____Spacing/Edge Distance

Comments:

SCREENING EVALUATION: LOAD PATH

Noted Load Path Concerns:

____Connections to Components   ____Missing or Loose Hardware
____Support Members   ____Other Concerns

Comments:

SCREENING EVALUATION:
STRUCTURAL INTEGRITY/EQUIPMENT SPECIFIC

Noted Structural Integrity/Equipment Specific Concerns:

____Maintenance   ____Functionality
____Brittle Material   ____Ground Failure
____Corrosion   ____Lateral Load
                                   ____Other Concerns

Comments:

SCREENING EVALUATION: SYSTEMS INTERACTION

Noted Interaction Concerns:

____Failure and Falling   ____Differential Displacement
____Proximity and Impact   ____Spray/Flood/Fire

Comments:

ADDITIONAL NOTES

SIGNATURES

Name: _____   Date: _____

Name: _____   Date: _____

Figure 6-1. This example walkdown sheet shows how a simple checklist can remind the engineer of what to evaluate for a given piece of equipment.

6.4 SYSTEM CONSIDERATIONS

In most practical applications, walkdowns are performed on a component basis. Individual items (e.g., equipment items, vessels, specific pipelines, or piping runs) are generally identified as important because of hazardous materials, fire potential, potential interaction with other components, etc., and are evaluated by the seismic walkdown team. Often, these evaluations are performed completely independently of any process hazards analyses (e.g., "What-Ifs," HAZOPS, etc.) that may have initiated the seismic evaluations to be performed. In those situations, the walkdown team must interact with teams performing process hazards analyses. A constructive information interchange greatly enhances the efficiency and potential benefits from the review.

The seismic review team should be able to describe to owners and process safety engineers some of the general effects that can be expected in a scenario earthquake. For example,

(a) The entire facility will be shaken simultaneously without prior warning.

(b) The shaking may last 10 s or longer. Very large magnitude earthquakes (greater than magnitude 8) have caused shaking lasting on the order of 60 s or more.

(c) Off-site power and water will likely be lost.

(d) Several systems may be lost at the same time, such as phones, water, etc., and for long periods of time.

(e) Underground piping may break.

(f) Certain vulnerable equipment items and piping systems may be damaged and unable to function.

(g) Off-site emergency services may not be available due to infrastructure problems (bridge or highway damage) or due to their required use in the general community.

(h) Concerns about personal or family safety may be the priority of operator personnel, and operator action may not be a viable method of mitigating damage.

If the review is part of a hazard analysis, all of these issues may affect the hazard analysis results and should be considered by the process safety team. The walkdown engineer should critically review what assumptions have been made in the process hazards analyses that might be inappropriate regarding post-earthquake capabilities of facilities and equipment.

The walkdown engineer should interact with operators, owners, process safety engineers, and other available specialists regarding consequences of damage. For example, a civil/seismic engineer might assume that the highest consequences of failed process piping are associated with pipes carrying the most toxic material. In reality, other considerations, such as whether the system will continue to feed material through the line, or whether the pressure drop will shut down production of the material, may be more significant factors in prioritizing the hazards.

This point is made to alert walkdown teams to be consistent in evaluations, calibrating judgment based on understanding of the systems, not on unfounded assumptions. It is also made to emphasize to the civil/seismic engineer that not all potential damage needs to be mitigated. The walkdown engineer will likely identify several concerns that are determined to lack significant safety or economic implications—a decision that will be made by others with input from the walkdown team. At that point, decisions regarding upgrades should be made by the owner on a cost-benefit basis.

6.4.1 Emergency Systems

During a damaging earthquake, off-site utilities will very likely be disrupted and potentially lost for long periods of time. If the scope of the review has been limited to equipment and piping with hazardous materials, the civil/seismic engineers should question the process safety engineers to determine their assumptions on the availability of off-site utilities following an earthquake. Including items such as backup power supplies and water storage tanks in the seismic evaluation may be prudent.

The walkdown team should also question whether other emergency systems are being counted on to function during a large earthquake to mitigate damage. In particular, whether the fire protection system, telecommunications systems, and containment systems will be required to perform active functions after the earthquake should be determined.

6.5 EVALUATION OF COMPONENTS

The following sections aim to give guidance for performing walkdowns of typical components in a petrochemical or other industrial facility. Note that this guidance is not intended to be all inclusive; listing all of the possible situations that an engineer might encounter is impossible. Rather, the following issues represent those that have been identified in the investigations of equipment performance in earthquakes throughout the world over the last two decades. In addition, the following incorporates the in-plant experience of engineers who have performed seismic walkdowns of petrochemical and other industrial facilities, primarily in California. In all cases, the walkdown engineer must use his or her common sense and fundamental principles of engineering mechanics, as necessary, in identifying potential seismic vulnerabilities.

6.5.1 Major Considerations

Several major considerations should influence the focal points of the walkdown investigation and the relative effort spent on various aspects. Examples include

(a) Level of ground shaking hazard: In areas of lower seismicity, major structures and vessels may be designed for sufficient lateral load capacity resulting from design for other criteria, such as wind. However, displacement-induced damage can occur at low levels of shaking.

(b) Severity of other hazards (faulting, soil failure, and landslides): Known faults near the site should alert the walkdown team to be on the lookout for situations where imposed displacement could cause damage, such as buried piping, or equipment supported on different structural systems. Where known faults run through the site, a walkdown evaluation may need to be supplemented with additional geotechnical or other investigations. Soil failure such as liquefaction could also greatly affect the severity of damage.

(c) Vintage of the facility and applicable codes at the time of construction: Applicable codes and seismic design methods may have changed considerably since a unit was designed. In particular, reinforced concrete design codes changed in the early 1970s, adding detailing requirements that would ensure ductile behavior in an overload condition. Evaluating overall structural capacity in older units should be emphasized more than in newer units. In addition, engineers evaluating older facilities should be more alert for existing damage, such as dents in structural members, damaged concrete, corrosion, etc.

(d) Overall quality of maintenance: Where overall maintenance appears to be poor or inconsistent, walkdown teams should be alert for missing nuts and bolts, unrepaired damage, field modifications, etc., especially in the primary load path and connections (Figure 6-2). Significant deterioration may also exist in structural members and their connections. The team should be on

Figure 6-2. Example of a significant modification to the load path.

Figure 6-3. Concrete beam cracking from corrosion of reinforcing steel.

the lookout for signs of active corrosion that may be somewhat hidden under fireproofing or insulation (Figures 6-3 and 6-4).

(e) Additions or modifications to structures since original design: Over the years, additional equipment may have been added to a structure, or the original structure may have been expanded horizontally or vertically. An overall seismic evaluation of the structure may not have been performed at the time of a past modification. A review of the existing drawings can provide valuable information regarding past modifications and provide information regarding any structural upgrades at the time of the additions or modifications.

(f) Priorities based on process safety considerations, pollution, regulatory needs, etc.: The process safety engineers and owners should identify to the walkdown team which components may warrant a more thorough initial review due to safety, pollution, or economic consequences of damage.

6.5.2 Other General Considerations

The following issues are common to petrochemical and other industrial facilities. Additional guidance is provided for several of these issues, as appropriate, in the discussion of specific equipment items.

6.5.2.1 Anchorage

Displacement resulting from inadequate or missing anchorage has probably been the most common source of damage to equipment in past earthquakes. Among the specific details noted to have caused problems are the following:

(a) Vibration isolators: Rotating equipment is often isolated from its support-ing structure by the use of elastomeric pads or springs that do not transmit

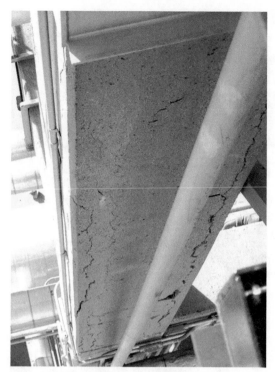

Figure 6-4. Concrete column cracking and spalling.

the vibrational loads from the equipment to the structure. Those isolators, which may appear on a quick visual review to be designed for lateral loads, often lack the strength and ductility to resist seismic loads without failing (Figure 6-5). One potential cause of failure is when the isolators are made of nonductile material, such as cast iron, which may fracture rather than deform under earthquake loading. Another potential cause of damage is a lack of vertical uplift restraints. The equipment may bounce out of the support and thereby lose its effective lateral support. A good detail on a vibration isolator commonly includes "bumpers" in the lateral direction (Figure 6-6) or stops that will not prevent motion but will limit the displacement to tolerable levels. A good detail also includes uplift restraints. In general, any vibration-isolated equipment should be carefully evaluated. Isolators themselves should be evaluated, as should the consequences of large displacements that may occur should that isolator fail.

(b) Welds: Nonductile failure may occur in situations where welds are overstressed. Situations of concern that might be identified during a walkdown evaluation include

- Corroded welds, a concern that should be evaluated wherever standing water accumulates or welds are constantly exposed to water;

Figure 6-5. This is an example of a failed vibration isolator. In this case, the spring broke. Other times, the mounting may be made of brittle material, such as cast iron and may fracture.

Figure 6-6. Bumpers should be included to limit the movement of vibration-isolated equipment. Uplift restraints should also be provided.

Figure 6-7. Example of a poor-quality weld, in this case a tank base plate to checker plate decking. Where welds may be difficult, the walkdown team should evaluate the quality closely.

- Potentially undersized welds;
- Situations where good-quality welds may be difficult to install, such as a weld to checker plate that may not be of high quality (Figure 6-7);
- Welds where they should not be (e.g., anchor bolts welded to their "seats" on tanks); and
- Welds over shim plates (Figure 6-8), an especially difficult detail to detect in a field review and unlikely to show in drawings.

(c) Bolted anchorages: Cast-in-place, grouted-in-place, and expansion anchor bolts may fail not only due to lack of strength, but due to details such as inadequate edge distance or cracks in the concrete (Figure 6-9). Factors such as spacing of the bolts may reduce the capacity of bolted anchorages due to overlapping shear cones. Spacing and edge distance may cause a reduction in capacity if distances between bolts or bolts and edges are less than 10 times the bolt diameter. Special attention should be given to inspection (e.g., bolt tightness checks) of grouted-in-place and expansion anchors, as their capacities are very sensitive to proper installation.

6.5.2.2 Load Path

An inadequate load path can lead to damage or even failure during a seismic event. The following load path issues should be looked for during a walkdown evaluation of an existing facility.

- The load path of the major equipment masses down to the foundation should always be visually reviewed.

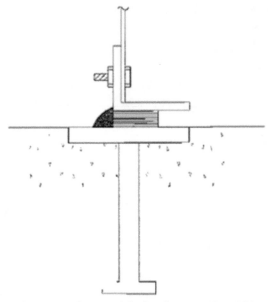

Figure 6-8. Shim plates are often used for leveling vessels. Welds over these shims may have severely reduced capacity. This poor detail is difficult to spot in drawings or in the site investigation. In this case, the walkdown team should notice the apparent thickness of the base plate.

Figure 6-9. Large cracks within 10-bolt diameters could reduce the tension capacity of an anchor bolt. Cracks such as these through the anchor bolts could significantly reduce their tensile and shear capacity.

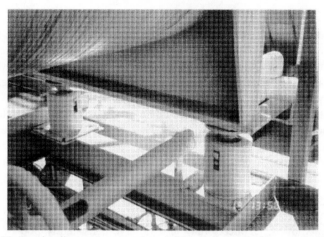

Figure 6-10. Load cells are commonly used in the support for vessels. These may lack the capability to resist lateral loads and should be carefully examined.

- The engineer should be aware of unusual cutouts or modifications. Where those cutouts are obviously field modifications, the engineer should assess whether lateral capacity is significantly reduced. Cutouts are common in support saddles, especially for horizontal vessels and heat exchangers where the anchor bolts on the piers are not correctly aligned with the predrilled holes in the steel saddles. They may also be found in vessel support skirts, when piping does not align with prefabricated openings.
- Load cells within the load path (Figure 6-10) may not be capable of resisting lateral loads. These are typically used for vessels.
- A common problem is missing nuts and bolts on connections. This situation is especially prevalent where structural members must be removed to provide access for regular maintenance of equipment.
- Attention must be given to cases where large eccentricities exist. This often occurs as a result of field modifications that have no engineering basis, such as a shifting of braces to allow for clearance for piping or other equipment. Eccentricities, if significant, can induce significant bending on structural members, such as columns, for which these members are not designed.

6.5.2.3 Maintenance

During walkdowns of facilities, engineers routinely observe conditions that compromise the seismic structural integrity of equipment yet could be easily taken care of through routine maintenance. This occurs even in otherwise well-maintained facilities. As noted above, this is primarily a load path/anchorage issue related to missing or damaged hardware.

6.5.2.4 Corrosion

The walkdown team should always be on the lookout for areas where corrosion may occur. The concern is not surface rusting, but a loss of structural strength that may be indicated by thinning, pitting, or flaking. Areas especially vulnerable would be where especially corrosive materials such as acid are present and where water may accumulate.

Another area where corrosion may be a problem is where concrete cover is spalled and the reinforcement is exposed. This is generally a matter of maintenance.

Because corrosion is often a high-priority general concern for a facility, the plant may have a corrosion group that can assist in identifying potential problem areas.

6.5.2.5 Construction/Installation Quality

During walkdowns, engineers may also observe consistently poor installation practices. This may be evident in welds, or in expansion anchor bolt installations. For example, expansion anchors may not be able to attain their tensile design capacity if embedment is inadequate. This could result from the use of shim plates or large grout pads. Long studs protruding above the concrete surface, or exposure of part of a shell insert, may indicate such problems. Other examples of installation concerns would include connections, such as cotter pins not installed properly or fasteners missing positive locking devices for vibrating equipment.

6.5.2.6 Seismic Interaction

Seismic interaction refers to damage to a system or piece of equipment due to impact with or movement of another piece of equipment, system, structure, storage cabinets, furniture, etc. This is a particular aspect where the walkdown evaluation is the best way to identify potential interactions.

For the purposes of this document, seismic interactions are divided into four primary categories:

1. Proximity and impact occur when clearance between two items is not adequate to prevent impact. This may occur from sliding of unanchored equipment; swinging of rod-hung piping, ductwork, or cable trays; or cantilevered deflection of electrical cabinets causing pounding with adjacent cabinets, walls, or structural members. Another example is the hazard of walkway platforms with sharp-ended beam supports, if the walkways are capable of movement relative to the tank such that the tank wall can be punctured.

2. Structural failure and falling usually occur when inadequately restrained components fall from above, impacting the equipment in question. This can also occur when a large item nearby structurally fails and impacts upon the equipment being evaluated by the walkdown engineer.

3. Differential displacement is a particular concern for the walkdown team wherever different structural systems support items. The engineer should be aware of potential situations where the different systems can displace, such

that connecting piping, ducts, conduit, tubing, and so on lack adequate flexibility to survive the motion. Flexibility is the key feature for resisting damage. This is also a specific concern when different foundations act as supports for one piece of equipment, or where equipment is unanchored.

4. Waterspray and flooding may be of concern inside buildings, where damage to sprinkler systems could affect the operability of electrical equipment. Of particular concern are sprinkler heads on fire piping that might impact hard structural members and open, spraying water on electrical equipment below.

6.5.2.7 Process Changes

Process changes may result in operating conditions that have been changed from the design conditions without consideration of the effect on seismic vulnerability. The walkdown team should make appropriate process engineers and operators aware of this possibility and investigate where such changes may have occurred. Examples include increased operating levels in vessels, larger operating equipment, or increased product density.

6.5.2.8 Inadvertent Relay Actuation / Functionality

The action of electromechanical relays during an earthquake may affect the functionality of equipment. This cannot be evaluated during a walkdown and requires a system evaluation. Walkdown engineers should be aware of the potential for relays chattering, tripping, or changing state due to earthquake shaking and that certain types of relays, such as switches utilizing mercury-filled vacuum tubes, have been demonstrated to be vibration sensitive. Walkdown engineers should notify process engineers and operators of this possibility and investigate further when this condition is a concern.

6.5.2.9 Ground Failure

Ground failure has led to severe damage and collapse of many otherwise well-designed structures and equipment in earthquakes.

(a) Walkdown engineers should always review soil reports to identify the potential for faulting, liquefaction, and settlement. Special caution should be exercised whenever one or more of these situations are encountered, because even well-engineered structures and equipment can be severely damaged. Walkdown evaluations generally must be supplemented with additional studies whenever these types of ground failure are possible.

(b) Steep slopes with potential stability problems should be identified, and if questionable, slope-stability studies should be recommended.

6.5.2.10 Design Interface

In general, the walkdown team should always be alert when investigating areas of design interface, where connecting structural elements may have been designed by different engineering groups. Examples include vessels on supports, where the vessel and saddle design may be by the vendor, while the attachment to the support and the support itself may be by others; or finfan units, where the pipeways may be designed by one group and the finfan and support framing by the vendor.

In these situations, the walkdown team should be alert for inconsistent design practices, an indication that one or more elements may not have been properly designed for seismic loads.

6.5.3 Evaluation of Specific Components

The following sections list considerations in evaluating several specific types of equipment and structures found in petrochemical and other industrial facilities. Again, the following guidance is not all inclusive and cannot possibly address all situations that might be encountered. It simply indicates observed causes of damage and observed vulnerabilities from past walkdowns.

6.5.3.1 Mechanical Equipment

Many types of equipment consist of complex assemblies of mechanical and electrical parts that are typically manufactured in an industrial process that produces similar or even identical items. Such equipment may include manufacturer's catalog items and are often designed by empirical (trial and error) methods for functional and transportation loads. One characteristic of this equipment is that it may be inherently capable of surviving strong motion earthquakes without significant loss of function. Equipment that may fit into this category include most air handlers; compressors; pumps; motors; engines; generators; valves; pneumatic, hydraulic, and motor operators; fans; chillers; evaporators; and condensers.

The engineer performing the walkdowns should review each of these items looking for possible fragile parts of the components and for specific configurations and details that have been shown to be potential problems, such as the following:

(a) Anchorage of the unit: In particular, vibration isolators may be found on air handlers, compressors, generators, and small pumps.

(b) Compressors, generators, and pumps: These items may have engines and motors located on separate skids or foundations. In those cases, the units should be investigated for potential damage due to differential displacement, such as binding of a shaft. This is a functionality issue and may not be an issue for toxic material, flammable material, or product releases.

(c) Attached piping: This must be flexible enough to withstand differential motion between its two anchor points. This may be particularly important when it is attached to unanchored equipment or tanks.

(d) Valves: Valves are generally very rugged seismically. However, cases exist where they have been damaged due to impact of the operator with a structural member. This is a special concern when the valve yoke is constructed of cast iron. The walkdown team should also investigate those situations where a valve and its operator are independently supported on different structural systems.

(e) In-line components: Additional investigation will also be appropriate where in-line components are large relative to the size of the piping.

(f) Nonductile materials, such as PVC or cast iron.

6.5.3.2 Electrical Equipment

Like mechanical equipment, much electrical equipment is designed and manufactured for functional and transportation loads and has characteristics of ample construction that make it inherently able to survive strong earthquakes. Well-anchored electrical equipment has typically performed well in past earthquakes. This includes motor control centers, low and medium voltage switchgear, transformers with anchored internal coils, inverters, battery chargers, batteries, and distribution panels.

When electrical equipment is required to function during or after an earthquake, the walkdown engineer should review each of these equipment items, looking for potential vulnerabilities. Examples of concerns for electrical equipment items include the following:

(a) The engineer should verify anchorage of electrical equipment. The anchorage may be bolts to concrete, fillet welds to embedded steel, or plug welds to embedded steel. Of particular concern are plug welds, which will not have the same capacity in tension as they will in shear. If a potential for overturning is present, the plug welds should be checked with a reduced capacity. To check weld capacity, use 25% of the capacity of an equivalent fillet weld around the perimeter of the hole.

(b) Cabinets that are not bolted together may pound against each other. This is a particular concern if the cabinets need to function after an earthquake and trip-sensitive devices are present, such as switches and relays inside the cabinet.

(c) Cabinets that are adjacent to structural columns or walls may experience a cantilever deflection, causing an impact. Except for extremely flexible cabinets, this should not be an issue where more than approximately 1 in. (25 mm) clear space is available. Again, this is a concern where trip-sensitive devices are present. This is not an issue if the cabinet is stiff due to top bracing or rigidly supported conduit coming from the top of the cabinet.

(d) The front-to-back shear panels should not have unusually large cutouts in the side or near the bottom that may compromise the cabinet's structural integrity. This does not refer to manufacturer's installed and reinforced cutouts, or cutouts and doors in the front and back panels.

(e) Internal devices should be secured to the cabinet structure or internal framing. For example, bolts connecting transformer coils to the cabinet are sometimes removed after transporting the unit to a site and installing the equipment.

6.5.3.3 Battery Racks

Emergency battery racks are identified separately because they have failed several times in earthquakes, often leading to a lack of emergency power. Walkdown engineers should evaluate the following:

(a) The battery rack should be structurally sound and capable of resisting transverse and longitudinal loads.

(b) The batteries themselves should be restrained from falling off the rack. This is typically done by installing wraparound bracing around the batteries (Figure 6-11).

(c) Where batteries have gaps between them, some form of spacer should be present to prevent sliding of the battery and impact or damage to bus bars.

(d) Falling of overhead equipment should be avoided to prevent possible electrical shortcircuits or damage to the batteries. Of particular concern are fluorescent tubes in lights, which have been observed to fall from fixtures and drop to the floor if safety grating is not present. All emergency lights, horns, speakers, etc., in the vicinity of batteries should be looked at to ensure that they will not slide, fall, or otherwise move such that they can hit the batteries.

Figure 6-11. Batteries must be kept from sliding off the racks. This is commonly done with wraparound bracing, as illustrated here.

(e) If the engineer is reviewing the batteries to ensure their functionality after an earthquake, related electrical equipment should also be included in the review and given special attention (e.g., inverters, control panels, etc.).

6.5.3.4 Control and Instrumentation Equipment

Control panels should be viewed with the same concerns as electrical cabinets. Of particular concern will be the presence of trip-sensitive devices, such as relays. When relays are present, the walkdown engineers should question process safety personnel or operators as to whether those relays are required to function during or after an earthquake.

Instruments on racks generally perform well, provided that they are secured to the rack and the rack is anchored. Instrumentation such as thermocouples and gauges are not an issue except where they can be damaged by impact or excessive motion pulling out cables.

A few other issues particular to control and instrumentation equipment are as follows:

(a) Control panels often contain components on rollers or slides. These drawers may not have stops and have been observed to roll out and fall to the floor during an earthquake. The walkdown engineers should check for stops or other restraints on components on slides or rollers.

(b) Circuit cards often slide in and out of panels with no restraining devices. These circuit cards have been observed to slide out and fall to the floor during earthquakes. The engineer should look for restraining devices. Some cards are restrained by tight friction and do not have latches. Retrofits may be impractical.

(c) Control panels often have doors left open or unlatched. The swinging of doors and resulting impact may be a concern if trip-sensitive devices are present.

6.5.3.5 Pressure Vessels

Vertical pressure vessels are often mounted on steel skirts and anchored to a concrete foundation. The vessels themselves are usually designed for high pressure and perform well in earthquakes. Tall vessels and columns are also often designed for wind loads and have significant lateral capacity. Of special concern for vertical vessels are the following issues:

(a) Unreinforced cutouts in skirts: Of special concern are cutouts that appear to be field constructed.

(b) Flexibility of attached piping: Rigid, separately supported piping can experience damaged nozzles or failed pipes as a result of differential displacement.

(c) Strength and ductility of the anchorage, such as anchor bolts with no chairs.

Horizontal pressure vessels are typically supported on steel cradles that are anchored to concrete piers. Where thermal expansion can occur, one end will be fixed, with the other end using slotted holes to allow for axial thermal growth. As with vertical vessels, the vessels themselves are designed for pressure loads and would be expected to perform well in earthquakes. The following concerns have been noted for horizontal pressure vessels:

(a) Unusually tall piers may not be capable of resisting transverse or longitudinal loads.

(b) Narrow piers may have cast-in-place anchor bolts with edge distance problems.

(c) Walkdown teams should be particularly aware of field modifications in the supports, such as cutouts to modify alignment of the bolts and bolt holes.

(d) Lack of flexibility in the piping attached to tanks and vessels can lead to a failure of the connection during a seismic event.

(e) For stacked horizontal heat exchangers, the walkdown team should check connection bolts between the exchangers. Nuts or bolts are commonly not replaced after maintenance or a turnaround.

Pressurized spheres are typically supported on several legs that are evenly spaced around the circumference of the sphere. The legs may or may not be braced, typically with X-bracing between adjacent legs. Failures observed in past earthquakes have been caused by failure of the support system. In the 1952 Kern County Earthquake, failure of the support legs led to breaking of attached piping. The butane that escaped the spherical tank eventually ignited, causing explosions and fires. The structural adequacy of the legs and bracing should be the primary focus in evaluating spherical tanks. Drawings should be reviewed to ensure that legs are tied together with grade beams.

Another consideration for pressurized spheres is that the support legs are typically fireproofed, creating a particularly vulnerable location at the top of the support leg where the fireproofing terminates at the vessel wall. The corrosion under fireproofing may not be obvious, but it can result in a loss of strength for the support legs and leave the sphere vulnerable in a seismic event. This has resulted in collapses during hydrostatic testing.

Small tanks and pressure vessels on legs are found throughout petrochemical and other industrial facilities. The tanks and vessels themselves have not been observed to be significant problems in earthquakes. These areas should be emphasized:The structural adequacy of the support legs to resist overturning must be considered. These have failed in many instances in earthquakes;flexibility of attached piping should be checked, which is especially true if the piping is connected to unanchored equipment or vessels that can move.

6.5.3.6 Finfans

Finfans are air coolers typically mounted on top of pipeways. The following considerations should be given to the seismic evaluation of these units:

- The structural adequacy of the support framing and the attachment to the pipeway should be evaluated. Further investigation is warranted if the framing shows signs of distress, such as buckling or bowing of members, under operating conditions. Attachments should be investigated where they are eccentric, or do not frame into primary structural members.

- Coil bundles are removed at regular intervals for maintenance purposes. These units may be positively attached to the support frame at one location or on one end only. Reinstallation of the coils may be incomplete, with missing bolts, misalignments of structural framing, etc. The walkdown team should investigate those interfaces.

6.5.3.7 Heaters

Boiler and heater structures are generally thick-walled steel vessels, supported on several low concrete piers. They may also be cylindrical, supported on skirts similar to other horizontal pressure vessels. Horizontal heaters are almost always fixed at one pier with slotted bolt holes at other piers to allow for thermal expansion.

The main focus on a boiler structure will be the support system. Reinforced concrete piers should be checked for adequate strength and stiffness, so that they do not create a "soft story" effect. The reinforced concrete piers should also be checked that the piers were detailed and provided with adequate confinement ties, especially near the top of the piers where they confine anchorage.

The walkdown team should be aware of the possible presence of refractory brick inside a heater or boiler structure. This will add weight to the equipment and may be an additional source of internal damage to the equipment.

6.5.3.8 Support Frames and Pedestals

Many equipment items will be located above grade, supported on reinforced concrete or steel frames. These frames may be irregular and are typically open at one end to allow for installation and removal of the vessels.

The failures of freeway structures in the 1989 Loma Prieta and 1994 Northridge earthquakes illustrate the potential for catastrophic collapse of older reinforced concrete structures without adequate shear reinforcement and confinement. A common configuration in refineries is to install large vessels, elevated high above the ground on pedestal supports (Figure 6-12). The support system typically has no redundancy. Reinforced concrete frames and pedestals should be carefully evaluated, especially if constructed prior to the mid 1970s, when changes in building codes added ductile detailing requirements. The walkdown team should review available structural drawings to verify adequacy of concrete structures. In many cases, some form of numerical evaluation may be necessary to determine adequacy of the reinforced concrete structures.

The walkdown team should also be on the lookout for damaged and cracked concrete, especially where reinforcement is exposed. Steel reinforcement in cracked concrete could be susceptible to moisture and corrosion, with subsequent loss of strength.

Figure 6-12. Vertical vessels are often supported on each end by a single reinforced concrete pedestal. These vessels may be 30 ft (9 m) above the ground or more, creating a large moment on the pedestal. Drawings may need to be reviewed for this type of configuration. Special caution should be given to supports designed prior to building code changes in the mid 1970s.

6.5.3.9 Steel Frames

Steel frames have traditionally performed well during earthquakes. However, several surprises were noted in the 1994 Northridge Earthquake (EERI 1996). More than 100 steel frame buildings were damaged as a result of that earthquake. The damage was on moment-resisting frames and was, for the most part, concentrated in beam to column connections (cracks were found in welded beam flanges to the column) and, in some cases, in fracture of column base plates.

This is of particular concern to the structural community because these types of structures have been thought to be the most ductile and because of the potential implications for the vast number of existing steel frame structures. Note, however, that none of the damaged structures collapsed and most remained functional.

The walkdown engineer should pay additional attention to the connections in steel frame structures in light of recent experience.

The walkdown team should be aware of signs of distress due to dead load conditions or corrosion. Particular attention should be given to cases where drawing modifications (brace relocation/removal) may have taken place subsequent to the original design of the structure. These modifications may be related to piping clearances, etc., and may have taken place without the consultation of a structural engineer. Brace relocation and removal can result in load paths being different than originally designed, with resulting potential structural inadequacy.

Most often, the weakest links—where failures can be expected—are in the connections. The connections should be investigated for signs of insufficient stiffening, indirect load path, potential prying action, etc.

Finally, in cases where nonstructural elements are attached to structural frames (platforms, guard rails, etc.), the walkdown engineer should check that these do not act to significantly alter the structure's dynamic response in an earthquake or to change the primary load path.

6.5.3.10 Buildings

This document will not detail the evaluation of existing buildings. That topic is covered in great detail in documents such as ASCE 41 (2013).

The engineer should be cognizant of the types of building construction that have a higher potential for complete or partial structural collapse. While minor damage or even collapse of some buildings may be acceptable, suspect buildings should be identified to the process safety group. The following items are not uncommon in petrochemical facilities and should be evaluated by the walkdown team:

(a) Unreinforced masonry buildings are susceptible to partial or complete collapse.

(b) Masonry infill in walls is susceptible to collapse.

(c) Tilt-up buildings are susceptible to partial collapse if they are poorly tied together.

(d) Structures with vertical or plan irregularities may perform poorly in an earthquake.

(e) Reinforced concrete frame buildings lacking ductile details may be susceptible to failure.

Attention should be given to the level of functionality required of specific buildings following an earthquake. For example, when evaluating a firehouse, the walkdown engineer should make an evaluation to ensure that the operability of the facility remains intact, such as the ability to open doors for fire engines to exit.

6.5.3.11 Stacks

Although unreinforced concrete stacks and chimneys have failed in earthquakes and tall steel stacks have buckled, stacks generally perform adequately in earthquakes. The walkdown engineer should be aware of significant changes in building codes over the years relative to tall, flexible structures, such as stacks. For example, a stack designed in a high seismic zone in the 1990s may have required a minimum lateral force of 0.20g. Using the current code, that same stack may require higher loads. However, the same structure may have been designed to 1/3 that load in the early 1960s. This discrepancy may be a special concern in evaluating a spread footing for overturning, as the footing may appear to be underdesigned. However, because of the cyclic nature of dynamic loading, foundations have not been observed to suffer gross overloading such that a stack could tip over in an earthquake. This should not be a concern unless soils are liquefiable or very weak.

Note that past experience shows that other structures attached to furnace stacks may be overstressed, such as trusses supporting ducts feeding to chimneys high above the ground. The walkdown team should be alerted to look for situations where flexible or sliding type connections are not present, as these connections tend to preclude significant interaction between stacks and other structures.

6.5.3.12 Gas Cylinders

Chemical releases have occurred in earthquakes due to gas cylinders toppling and rolling. Toppling of cylinders also has the potential to create a missile hazard. The following guidance applies to gas storage cylinders:

(a) Gas cylinders are often chained to prevent falling during normal operations. However, single chains do not necessarily prevent falling. Chains supporting both the upper and lower portion of the cylinder are necessary to prevent the cylinders from falling and sliding or rolling.

(b) Engineer should ensure that cylinders that appear to be secure are, in fact, secured to structural members. This is particularly true inside buildings, where straps may be tied to nonstructural elements.

(c) Horizontal cylinders should be secured so that they cannot roll and fall off of supports. Figure 6-13 shows an example of a good detail, where removable hold down bolts are used to secure chlorine cylinders. Where hold-down devices are present, but appear to be ignored or incorrectly used, the walkdown engineers should question operators regarding typical procedures for restraining cylinders.

6.5.3.13 Chemical Storage Areas

Chemicals have fallen from shelves, reacted with other spilled chemicals, and caused fires in past earthquakes and can cause concern for explosions. Walkdown engineers should investigate the following issues:

(a) Engineers should determine whether cabinets are secured to prevent falling.

(b) Engineers should determine whether restraints prevent contents from being spilled off of shelves.

(c) Where a potential for chemical spill is identified, the engineer should question the operators as to the potential consequences of spills. Note that several mitigation methods may be available, such as physical restraints or separation of incompatible chemicals.

Figure 6-13. Cylinders that are removed when empty often sit unrestrained on vertical saddle-shaped supports. In this photo, a simple hold-down bolt prevents motion while the cylinder is in use. The presence of a mechanism like this can be judged to reasonably prevent overturning or sliding of the cylinder.

6.5.3.14 Piping

Process piping runs throughout petrochemical and other industrial facilities, running directly between pieces of equipment, or often supported on overhead pipeways. In many situations, the piping of interest will be particular lines containing specified quantities of hazardous material. Several difficulties in evaluating individual piping runs include

(a) With several lines on congested overhead racks, initially locating and then following specific piping runs becomes extremely difficult.

(b) Piping may be difficult to locate, even with piping and instrumentation diagrams (P&IDs) or flow diagrams. These drawings do not adequately represent physical locations or distances of piping runs. Layout drawings are usually not available.

(c) The walkdown engineer may have difficulty in identifying the boundaries of concern, especially considering bypass lines, injection lines, valves, etc.

Given these difficulties, evaluating all piping in a given area is often more practical and efficient. That method does require additional interaction with the process safety engineers to determine the consequences of postulated damage.

In evaluating process piping, stresses due to inertial loads are not the primary consideration, except in certain circumstances. Rather, the focus is on vulnerable details, fragile attachments, and connections that may experience severe displacements. The following observations should be considered when performing the evaluation:

(a) Welded steel piping generally performs well in earthquakes and is typically not susceptible to damage from inertial loading. Inertial loads may be a concern for nonductile materials, such as cast iron or PVC. Note that materials such as cast iron are probably used because of their noncorrosive properties, and changing the material on the pipeline is not a reasonable option. In these cases stress analysis may be necessary.

(b) Lack of lateral supports will not necessarily lead to failure of the pipe. Piping often spans very long distances without lateral supports with no damage or failure. Rather, the emphasis should be on ensuring that the piping system will not lose vertical support. For example, the walkdown engineers should look for situations where the pipeline is near the edge of the support and could slide off (Figure 6-14).

(c) Where lateral support is lacking, making the pipe very flexible, the walk-down team should also be aware of situations where the pipeline may be effectively "anchored" by a rigid connection of a branch line (Figure 6-15). If the branch line is significantly smaller than the header (less than one half the diameter), the pipe could be overstressed at the connection.

(d) The walkdown team should look for situations where vulnerable appurte-nances or portions of a pipeline could be impacted and damaged due to

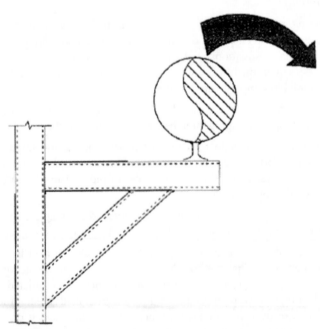

Figure 6-14. Pipes near the edge of a support without lateral stops may slide off. Loss of vertical support is the primary concern in these situations, not the length of span without lateral support.

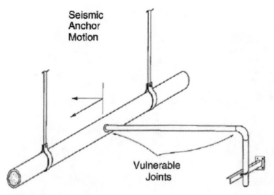

Figure 6-15. Rigid or restrained branch lines may be vulnerable to unrestrained motion of the header.

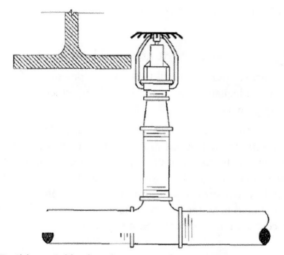

Figure 6-16. Fusible sprinkler heads are sensitive to impact.

motion of the pipeline or the impacting object. Examples include drain taps or sampling lines that could be impacted by valve diaphragms on adjacent lines. Of special concern are sprinkler heads on fire lines that might impact hard sharp structural members (Figure 6-16). This has occurred several times in past earthquakes.

(e) Mechanical couplings could fail due to excessive displacement or impact.

(f) Rod-hung systems, particularly fire protection systems within buildings, can fail and fall when supported by short rod hangers configured such that moments can develop at the top connection. This is a low-cycle fatigue failure.

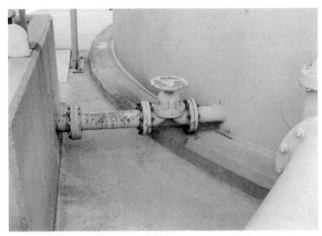

Figure 6-17. Piping damage is common where it is overrestrained at unanchored tanks, which can uplift significantly during earthquakes.

(g) Buried piping could fail where ground failure can occur.

(h) Piping attached to unanchored tanks or equipment can fail if flexibility in the pipe is inadequate to withstand large displacements in the event of tank uplift (Figure 6-17).

(i) In many instances, a relatively short span of large-diameter (8 in. or more) pipe connects elevated vessels supported by independent structural systems such as concrete frames. The stiff pipe may affect the interaction of the two structures and can result in overstress at the flanges or elbow areas, particularly when stress concentrations are accounted for. The walkdown engineer should take this into consideration and if in doubt recommend a proper coupled analysis of the entire system.

6.5.3.15 Cable Trays and Conduit

Cable trays and conduit are generally very rugged seismically. Note that damage to the trays and conduit themselves do not necessarily imply damage to the cables contained within them. The most important consideration for cable trays and conduit is to ensure that vertical support capacity will be maintained during a seismic event.

6.5.3.16 Ductwork

Ducting is often used in petrochemical and other industrial facilities to transport hazardous gasses and may be identified by the process safety engineers as requiring review. Duct construction can range from thin-gauge sheet metal to steel pipe sections. Sheet metal ducting has failed during past earthquakes in certain circumstances. Failure of sheet metal ducting generally is caused by corrosion, poor connections, loss of structural support, or differential displacement concerns. Large-diameter, thick ducting, such as plate steel construction, should follow the general considerations presented earlier on piping, especially concentrating on overall support systems and adequate flexibility to accommodate differential motions.

6.5.3.17 Air Tubing

Tubing is very ductile and is usually installed with substantial flexibility. However, tubing can be damaged by large displacements or impacts, and the consequences of loss of air to instruments should be considered.

6.5.3.18 Substations

Substation equipment may be especially susceptible to damage in earthquakes. Ceramic insulators often fail. Transformers that are unanchored, or lightly anchored with friction clips, may slide and damage connections.

Figure 6-18. Eccentrically loaded connection in a wood cooling tower. The cross member above is a retrofit.

6.5.3.19 Cooling Towers

Wood cooling towers have generally not performed well in earthquakes. The following have been the primary causes of damage:

- The primary cause of damage to wood cooling towers in earthquakes has been a deteriorated condition prior to the earthquake.

- Cooling towers have also been damaged due to poor structural configurations at connections. For example, Figures 6-18 and 6-19 show an eccentric joint that failed during an earthquake, emphasizing that the engineer should focus on details first, rather than strength, when performing walkdown evaluations.

6.5.3.20 Platforms / Walkways

A refinery or other industrial facility has many elevated platforms and walkways. These platforms often surround tall vessels and are supported on a structural system that may not be directly connected with the vessels. In cases like these, differential displacement is the primary concern, with the potential for damage in the following ways:

Figure 6-19. Identical joint in the same wood cooling tower as shown in Figure 6-18.

- Pounding between the vessel and catwalks, potentially resulting in either rupture of the vessel wall or sufficient damage to the platforms themselves that they are inaccessible in an emergency after the earthquake. The walkdown engineers would look for not only the proximity of the platforms to the vessels, but also for sharp ends of supporting beams that could increase the potential for damage.

- Walkways spanning different structures without adequate flexibility, such that differential motion could cause collapse and falling of the walkways.

6.6 LIMITATIONS

Because the walkdown review is intended to be a screening process, all concerned parties must understand that identifying all sources of seismic risk in such a mostly visual review is impossible. Items identified as potential hazards in a walkdown may survive an earthquake; likewise, items that appear to be adequate could fail for reasons not apparent during a visual or analytical review.

The engineer must always caution both owners and regulators that seismic risk can never be completely eliminated. Even with a more thorough review, some level of seismic risk will always exist, no matter how much time, effort, and money are spent evaluating and upgrading a facility. The risk can be better understood and reduced, but there are never any guarantees.

References

ASCE. 2013. *Seismic evaluation and retrofit of existing buildings.* ASCE 41. Reston, VA: ASCE.

CalARP. 2013. *Guidance for California accidental release prevention (CalARP) program seismic assessments.* San Diego: CalARP Program Seismic Guidance Committee.

EERI (Earthquake Engineering Research Institute). 1996. Vol. 2 of *Northridge Earthquake reconnaissance report.* Oakland, CA: EERI.

EPA. 1999. *Chemical accident prevention provisions.* 40 CFR Part 68. Washington, DC: USEPA.

OSHA. 2014. *Occupational safety and health administration standards.* 29 CFR 1910. Washington, DC: US Dept. of Labor.

CHAPTER 7

Design and Evaluation of Tanks at Grade

7.1 INTRODUCTION

This chapter provides the user with guidelines for designing new tanks and evaluating existing tanks for seismic loads. However, evaluation of existing tanks is emphasized. The information contained herein is mostly applicable to unanchored flat-bottomed storage tanks at grade in areas of higher seismicity, with requirements to be de-rated for lower seismic areas. Some guidance is also provided for anchored storage tanks.

Oil storage tanks are typically designed according to the industry standard API 650 (2014). Seismic design provision in Appendix E of API 650 first appeared in the 3rd revision of the 6th edition, dated October 15, 1979. Although the general theory was developed earlier, few tanks were designed for earthquakes before the code provisions came into effect. Since then, several experimental programs and numerous field observational data have suggested that API 650 Appendix E is conservative in most cases. The conservatism may be appropriate for design of new tanks, but not necessarily for assessment of existing tanks.

Because only recently constructed tanks have been designed to resist earthquakes, several potentially damageable tanks may be present in any given tank population. To limit a facility's exposure to earthquake damage, potentially damageable tanks should be identified, evaluated, and if required retrofitted.

Because seismic design provisions were only introduced into the API code in the late 1970s (Wozniak and Mitchell 1978), tanks that are older than this may be considered vulnerable to damage by large earthquakes. However, many liquid storage tanks built before these provisions were introduced have been able to withstand strong ground shaking with minimal or minor damage and were able to continue functioning.

This chapter discusses the steps involved in undertaking a seismic hazard mitigation program for existing tanks are discussed in this section, including

- Quantification of the site-specific seismic hazard;

- Walkthrough inspection to assess piping, stairway and walkway attachments, and other potential hazards;
- Analytical assessment of tanks to evaluate the potential for overturning and shell buckling; and
- Mitigation of seismic hazard. The most commonly used hazard mitigation measures include addition of flexibility to rigid attachments, reduction of safe operating height, and, as a last resort, anchorage of the tank.

7.2 PAST EARTHQUAKE PERFORMANCE OF FLAT-BOTTOMED TANKS

The following discussion on earthquake response of tanks is generally extracted from material presented in ASCE (1984) and Dowling and Summers (1993).

Flat-bottomed vertical liquid storage tanks have sometimes failed with loss of contents during strong earthquake shaking. In some instances, the failure of storage tanks had disastrous consequences. Some examples include fires causing extensive damage to oil refineries in the 1964 Niigata, Japan, Earthquake and the 1991 Costa Rica Earthquake; polluted waterways in the 1978 Sendai, Japan, Earthquake; fires and failure of numerous oil storage tanks in the 1964 Prince William Sound Earthquake in Alaska; and failure of numerous liquid storage tanks, both above and below ground, in the 1971 San Fernando, the 1980 Livermore, the 1985 Chilean, the 1991 Costa Rica, the 1992 Landers, and the 1999 Izmit earthquakes.

The response of unanchored tanks, in particular, during earthquakes is highly nonlinear and much more complex than implied in available design standards. The effect of seismic ground shaking is to generate an overturning force on the tank. This, in turn, causes a portion of the tank baseplate to lift up from the foundation. The weight of the fluid resting on the uplifted portion of the baseplate, together with the weight of the tank shell and roof, provides the restraining moment against further uplift. While uplift, in and of itself, may not cause serious damage, it can be accompanied by large deformations and major changes in the tank wall stresses. This is especially apparent when the seismic loading reverses and the (formerly) uplifted segment moves down impacting the ground and introducing high compression stresses into the tank shell. Tank uplift during earthquakes has been observed many times, but the amount of uplift is rarely recorded. A 100 ft (30.5 m) diameter, 30 ft (9.1 m) high tank uplifted 14 in. (350 mm) during the 1971 San Fernando Earthquake. During the 1989 Loma Prieta Earthquake, two 42 ft (12.8 m) diameter, 28 ft (8.5 m) high tanks uplifted 6 to 8 in. (150 to 200 mm); and a tank uplifted 18 in. (450 mm) in the 1964 Alaska Earthquake (EERL 1971; EERI 1990; National Research Council 1973). Whereas tank uplift is very common, sliding of tanks is rare and generally need not be considered a credible failure mode for storage tanks at grade.

In general, tanks, especially unanchored tanks, are particularly susceptible to damage during earthquakes. This is because all of the mass contributes to the overturning moment, but only a small portion of the mass contributes to the overturning resistance (the reason for this is that the contained fluid and the relatively flexible tank shell and bottom plate cannot transfer the lateral shear induced by the earthquake to the foundation). Some examples of tank damage that has occurred in past earthquakes include

(a) Buckling of the tank wall, known as "elephant foot" buckling. Essentially, this occurs because the vertical compressive stresses in the portion of the tank wall remaining in contact with the ground (i.e., diametrically opposite the uplifted portion) greatly increase when uplift occurs. More precisely, that portion of the tank shell is subjected to a biaxial state of stress, consisting of hoop tension and axial compression. In addition, the baseplate prevents the radial deformation that would normally occur under internal pressure. As a result, bending stresses are introduced into the shell wall, further increasing the tendency to buckle. The photograph in Figure 7-1 (taken after the 1992 Landers Earthquake in Southern California) shows a classical example of elephant foot buckling and accompanying failure of overconstrained piping. This buckling mode is normally associated with larger-diameter tanks having height to diameter (H/D) ratios of about 1 to 1.5.

Another common buckling mode normally associated with taller tanks having H/D ratios of about 2, is "diamond shape" buckling. Unlike elephant foot buckling, which is associated with an elastoplastic state of stress, diamond shape buckling is a purely elastic buckling.

Figure 7-1. Elephant foot buckling and failure of rigid piping. County of San Bernardino water storage tank, Landers, CA. Landers Earthquake, Richter magnitude 7.4, June 28, 1992.

A feature of tanks that experience either elephant foot or diamond shape buckling is that the buckled tank often does not rupture and continues to fulfill its function of containing fluid. However, all buckled sections of the tank should be replaced.

(b) During tank uplift, the baseplate may not be able to conform to the displaced shape of the tank and the weld between the baseplate and the tank wall may not be able to accommodate the tension stresses that develop as a result of the fluid hold-down forces that are mobilized to resist uplift. In this case, fracture at the junction between the baseplate and the shell wall may result. This was observed, in particular, in the Chilean Earthquake of 1985, where some tanks experienced failure of the baseplate or of the weld between the wall and the baseplate (EQE 1986). Good welding practices may preclude such failures, which can lead to rapid release of the tank contents; further, the differential pressures created within the tank by the rapidly evacuating fluid may also lead to damage to the upper shell courses and roof.

(c) Seismic shaking causes the surface of the tank fluid to slosh. If insufficient freeboard is provided to accommodate this sloshing, damage to the tank's floating roof or fixed roof, followed by spillage of fluid over the tank walls, may result. This type of damage is usually considered only minor, but may be important for some stored products.

(d) Support columns for fixed roof tanks have buckled in past earthquakes from the impact of sloshing fluid and their own inertial force in combination with the vertical loads they support. Collapse of several columns, while often not a catastrophic failure, can be a significant damage item.

(e) Breakage of piping connected to the tank as a result of relative movement between the tank and the nearest pipe support. This is one of the most prevalent causes of loss of contents from storage tanks during earthquakes. Failures of this type are typically due to inadequate flexibility in the piping system (termed "overconstrained piping") between the nozzle location at the tank shell and the adjacent pipe support. Failures have also occurred due to relative movement between two different tanks connected by rigid piping.

(f) Tearing of tank wall or tank bottom due to overconstrained stairways anchored at the foundation and tank shell.

(g) Tearing of tank wall owing to overconstrained walkways connecting two tanks experiencing differential movement.

(h) Anchored tanks, with nonductile connection details have performed poorly in past earthquakes. Nonductile connection details can lead to tearing of the tank's shell and release of its contents. Details that do not allow for full development of anchor bolts have led to anchor bolt pullout and failure of the tank.

(i) Tank failures have occurred as a result of severe distortion of the tank bottom at or near the tank side wall due to a soil failure. This failure may be associated with soil liquefaction, slope instability, excessive differential settlement, bearing failure, or washout due to pipe failure. A failure of this type occurred during the Miyagi Earthquake of 1978.

(j) Tensile hoop stresses due to shaking-induced pressures between the fluid and the tank walls can become large and can lead to splitting and leakage. This has been observed during earthquakes but appears to be a problem only for bolted or riveted tanks. For welded tanks, material ductility appears to accommodate these high hoop stresses.

Figures 7-2 through 7-6 present some examples of tank failures in earthquakes, including overturning of a tank during the 1991 Costa Rica Earthquake.

7.3 WALKTHROUGH INSPECTION

A walkthrough is an on-site visual screening, whereby as-installed components can be evaluated for seismic vulnerabilities. Chapter 6 discusses walkthroughs in general in more detail. The purpose of the walkthrough is to identify various seismically vulnerable details. In this chapter, walkthrough evaluation guidance is given specifically for tanks.

The principal feature that distinguishes the seismic response of unanchored tanks from that of anchored tanks is the large uplift displacements commonly observed around the edge of unanchored tanks. As described earlier, this uplift may induce large tension or compression forces and bending moments in the tank wall, baseplate, and at the intersection of the two. Such forces may lead to severe damage or failure of the tank. Furthermore, the adverse effects of excessive tank uplift can be greatly exacerbated by commonly encountered tank details. Walkthrough inspection of individual tanks or tank farms should focus on identifying such details. In many cases, while the tank itself may be found to be structurally adequate, retrofit of some of these seismically vulnerable details may be deemed necessary. The following material, which describes the walkthrough process, is taken from Dowling and Summers (1993) and Summers and Hults (1994).

The most frequently encountered hazardous details are listed below, together with appropriate retrofit recommendations, and illustrated in Figures 7-7 and 7-8.

(a) A common failure mode in tanks has been breakage of piping connected to a tank as a result of relative movement between the tank and the nearest pipe support. Alternatively, if the piping is stronger than the tank wall or baseplate to which it is connected, tearing of the wall or baseplate may result. Piping should not pass directly, with little or no flexibility, from the tank shell or tank bottom to the ground or to rigid concrete walls, basins, pumps rigidly fixed to the ground, etc. Failures of the type described above are typically caused by the details shown in Figures 7-7a through 7-7d.

Figure 7-2. Elephant foot buckling (top) and failure of rigid piping and tank wall failure at manhole access (bottom). Bighorn Desert View Water Agency water storage tank, Landers, CA. Landers Earthquake, Richter magnitude 7.4, June 28, 1992.

In the first three cases, additional piping flexibility should be provided by adding horizontal or vertical bends, or by installing a length of flexible piping. In the fourth case, piping should be rerouted to the center of the tank or, if the piping is flexible enough, the concrete basin may be extended beyond the pipe/tank connection.

Figure 7-3. Tank wall failure at tank manhole access, Bighorn Desert View Water Agency water storage tank, Landers, CA (top). Overall view of failed water storage tank, Landers, CA, County of San Bernardino, water storage tank, Landers, CA (bottom). Landers Earthquake, Richter magnitude 7.4, June 28, 1992.

(b) Similar failures have also occurred due to relative movement between two tanks connected by a rigid pipe, as shown in Figure 7-7e. Again, additional piping flexibility should be provided as described above.

(c) Partial loss of contents may result from the type of detail shown in Figure 7-7f, where a vertical pipe is rigidly connected to the ground or foundation and also supported rigidly along the wall of the tank. A detail offering a lesser level of risk, but present in many cases, is a tank wall support that consists of a large U-bolt that might appear to be capable of sliding up and down the pipe as the tank lifts and falls. However, the U-bolt could "bind" with the pipe, thereby also forming an essentially rigid

Figure 7-4. Overturned oil storage tank, RECOPE Oil Refinery, Costa Rica (top). Spilled oil and ruptured piping, RECOPE Oil Refinery, Costa Rica (bottom). Costa Rica Earthquake, Richter magnitude 7.4, April 22, 1991.

connection and leading to tearing of the tank wall. Any connection along the tank shell judged to be rigid should be replaced by a connection near the shell/roof intersection, coupled with sliding connections or "guides" along the shell wall. In many cases, simply loosening the U-bolts will suffice.

(d) Roof access is frequently facilitated by walkways spanning between the tanks. Figure 7-7g shows typical walkway arrangements. In both cases, relative movement between tanks may lead to rupture or tearing of the tank wall or roof. However, whereas the lower walkway arrangement in Figure 7-7g may lead to partial loss of tank contents, the upper walkway

Figure 7-5. Damaged roof, seals, and sloshing of oil over tank walls, RECOPE Oil Refinery, Costa Rica (top). Severe elephant foot buckling, Transmerquim Tank Farm, Costa Rica (bottom). Costa Rica Earthquake, Richter magnitude 7.4, April 22, 1991.

arrangement will, at worst, lead to damage to the walkway itself and/or to the roof; hence, no loss of contents will result. The distinction between the two arrangements is important because the lower walkway represents a concern that could result in release of product, whereas failure of the upper walkway will likely only lead to a release of fumes and a much lower level of

Figure 7-6. Damaged tanks because of fire after earthquake, 1999 Izmit, Turkey Earthquake.

economic loss. In either case, the required retrofit would take the form of increased walkway flexibility.

Elevated walkways also represent falling hazards for tanks located in plants with significant adjacent pedestrian traffic or sensitive equipment. Where necessary, walkways should be attached to the tanks by cables as a secondary means of support to prevent them from falling.

(e) Stairways should not be attached to both the tank shell and the foundation; see Figure 7-7h. However, only in thin-shelled tanks is such a detail likely to lead to failure of the tank wall and to loss of contents. Again, therefore, a distinction should be drawn between this case and the case of a thick-shelled tank, where such a detail would only result in damage to the stairway itself. In either case, the hazard may be eliminated by attaching the stairway to the tank shell or by eliminating the connection that prevents the stairway from displacing vertically.

(f) Tank piping is usually concentrated in one area, often with an access walkway spanning over the piping (Figure 7-7i). If the walkway is rigidly attached to the ground and if insufficient clearance is provided between the piping and walkway, then tank uplift may lead to impact between the piping and the walkway, resulting in damage to one or the other. However, only with small-diameter pipes or thin-shelled tanks is loss of tank contents likely to occur. Otherwise, damage is likely to be confined to the walkway itself. In both cases, the hazard may be mitigated by increasing the piping flexibility, attaching the walkway exclusively to the tank shell, or providing more piping clearance. Another potentially seismically vulnerable detail is the case of a walkway attached to both the tank shell and foundation, as in Item (e) above for stairways.

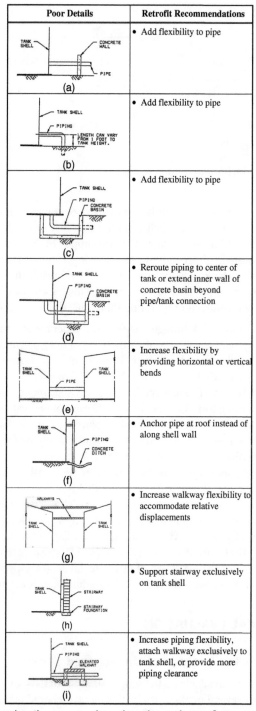

Poor Details	Retrofit Recommendations
(a)	• Add flexibility to pipe
(b)	• Add flexibility to pipe
(c)	• Add flexibility to pipe
(d)	• Reroute piping to center of tank or extend inner wall of concrete basin beyond pipe/tank connection
(e)	• Increase flexibility by providing horizontal or vertical bends
(f)	• Anchor pipe at roof instead of along shell wall
(g)	• Increase walkway flexibility to accommodate relative displacements
(h)	• Support stairway exclusively on tank shell
(i)	• Increase piping flexibility, attach walkway exclusively to tank shell, or provide more piping clearance

Figure 7-7. Poor details at unanchored tanks and retrofit recommendations.

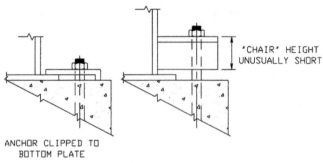

Figure 7-8. Poor anchorage details.

(g) Tanks anchored with anchor bolts having poor connection details may tear the bottom plate or tank shell resulting in a loss of product. Poor details (see Figure 7-8) include anchors that are clipped to the bottom plate, chairs that are unusually short and thus not allowing adequate transfer of forces in the bolt to the tank shell, or any detail that will result in the tearing of the tank shell before the anchor bolt yields. This hazard can be mitigated by replacing the connection with one that will exhibit more ductility.

When performing walkthrough inspections, experienced engineers familiar with seismic design and the effects of earthquakes should be consulted to answer questions regarding "how much flexibility is sufficient." The assumed value of tank uplift is critical to answering this question. Values of 6 to 8 in. (150 to 200 mm) have been common in the past. The first version of these Guidelines recommended using values on the order of 6 to 12 in. (150 mm to 300 mm) of vertical displacement and 4 to 8 in. (100 to 200 mm) of horizontal displacement in the zones of highest seismicity. Design values to be applied to the connecting piping for both horizontal and vertical movements are now given in ASCE 7-16, AWWA D100 (2011), and API 650. The values shown in these documents are not consistent with each other. API 650 treats the values shown in Table 15.7-1 of ASCE 7 as allowable stress-based values and therefore requires that these values be increased by a factor of 1.4 when strength-based values are required. API 650 amplifies the allowable stress-based displacements by a factor of 3, which is a slight rounding of the theoretical value of $1.4C_d$. AWWA D100 only deals with the allowable stress-based displacements. Actual expected values are a function of tank size, fill height, aspect ratio, and local seismicity and soil type. Section 7.4.5 provides additional discussion.

7.4 ANALYTICAL EVALUATION

7.4.1 Methods for Analysis of Unanchored Tanks

Extensive literature is available on the seismic analysis of flat-bottomed tanks. The literature includes several methods for analyzing unanchored tanks. Some of these analysis methods are

(a) API 650 Appendix E: This method is the standard for design of new tanks for the petrochemical industry.

(b) AWWA D100: This method is very similar to the API 650 method, and is used primarily for design of water storage tanks.

(c) "Earthquake Tank-Wall Stability of Unanchored Tanks" (Manos 1986): George Manos developed an alternative method of assessing tanks, which differs significantly from the previous two methods and is based on observed performance of tanks during past earthquakes.

(d) "Seismic Design of Storage Tanks" (Priestley et al. 1986): This method was proposed as the tank design code for New Zealand. It was not adopted at the time for steel tanks but was incorporated in NZS 3106 (1986), *Code of Practice for Concrete Structures for the Storage of Liquids*. Of all the methods listed, it is the most involved and has the most detail.

More discussion of the above assessment methodologies is given in subsequent sections.

7.4.2 Methods for Analysis of Anchored Tanks

Past performance of anchored tanks during earthquakes has indicated that such tanks generally experience very few problems. This is partially because very few tanks were anchored before the seismic provisions of API 650 were introduced in the late 1970s (Wozniak and Mitchell 1978), and those that were subsequently anchored were designed to those very same provisions.

Anchored tanks should be considered adequate unless anchorage details are judged to be capable of tearing the tank shell or bottom plate, causing loss of contents. Anchorage details should be assessed during walkthrough inspection to ensure that the load path is adequate for the hold-down forces developed in the anchor bolts to be transferred from the tank shell to the foundation.

If an anchored tank needs to be evaluated analytically, the methods in API 650 may be used. ASCE (1984) and Housner and Haroun (1980) also provide valuable background information.

7.4.3 Stability

API Evaluation Methodology

The seismic design methodology for welded steel storage tanks presented in API 650 Appendix E is based on the simplified procedure developed by Housner (1957). Wozniak and Mitchell (1978) describe details of the development of the API methodology. The procedure considers the overturning moment on the tank to be the sum of

(a) The overturning moment due to the tank shell and roof, together with a portion of the contents that moves in unison with the shell, acted on by a horizontal acceleration. The value of the acceleration is usually taken as the peak of the 5% damped site response spectrum, divided by a factor that accounts for the tank's ductility and reserve capacity. Alternatively, the code provides simplified formulas to determine the acceleration to be used. This is termed the impulsive component.

(b) The overturning moment due to that portion of the tank contents that moves in the first sloshing mode (i.e., independently of the tank shell), acted on by a horizontal acceleration equal to the 0.5% damped spectral acceleration corresponding to the period of that mode, again divided by a ductility/capacity factor. Alternatively, the code provides simplified formulas to determine the acceleration to be used. This is termed the convective component.

Resistance to the overturning moment is provided by the weight of the tank shell and roof and by the weight of a portion of the tank contents adjacent to the shell. The structural adequacy of the tank is determined by an "anchorage ratio," which is a measure of the ratio of the overturning moment to the resisting moment, with due consideration of vertical seismic motions. The anchorage ratio is defined in API 650 and must not exceed 1.54 (no metric equivalent is provided). Further, API 650 provides a methodology for calculating the compressive stress at the bottom of the tank shell together with the maximum allowable value of shell compression; the latter corresponds to approximately one-third to one-half of the theoretical buckling stress of a uniformly compressed perfect cylinder. If the anchorage ratio exceeds 1.54 or the compressive stress exceeds the allowable value, redesign of the tank or reduction of the liquid height is necessary. Anchorage ratios greater than 0.785, but less than 1.54, indicate that the tank may uplift but is stable. Once again, shell compression needs to be checked.

Extensive experimental studies and observations during past earthquakes have demonstrated that the radial length of uplifted bottom plate, and hence, the actual liquid weight resistance that is mobilized during an earthquake is underestimated by the API uplift model. The reasons for this are that the API model does not account either for the in-plane stress in the bottom plate, or for the dynamic nature of the tank response. The model also calculates a very narrow compression zone at the toe of the tank, thus leading to large compressive stresses in the tank shell for relatively low overturning moments. Finally, the API approach does not account for the effect of foundation flexibility on the tank wall axial membrane stress distribution.

Although the API methodology is known to be somewhat conservative and condemning existing tanks for failing to meet the API criteria may not be deemed appropriate, the criteria are the basis of the current seismic design practice and serve as a good benchmark. Large exceedance of specific provisions should be taken as an indication that retrofits may be necessary.

Alternative Methodology

Several alternatives to the API methodology were described previously; one that might be considered for use in evaluation of existing tanks is a modified version (Dowling and Summers 1993) of a method developed by George Manos (Manos 1986) presented herein. Manos's method is based on experimental studies and on observed behavior of unanchored tanks during past earthquakes. Instead of trying to model the complex uplifting plate behavior, Manos assumes a stress distribution at which the shell will buckle and solves for the resisting moment produced by the sum of the stresses. This resisting moment can then be compared with the overturning moment and the resisting acceleration solved for.

The assumed compressive stress is maximum at the toe (opposite the point of maximum uplift) and decreases to zero at an empirically determined distance from the toe. The maximum compressive stress is limited to 75% of the critical buckling stress of a uniformly compressed perfect cylinder. The model ignores the effect of hoop stress on buckling stress, but correlation of the method with actual test results validates the results obtained. Integration of the assumed axial stress yields a resultant compressive force that must be balanced by an equal tensile force due to the weight of the fluid resting on the uplifted portion of the bottom plate. Using an empirical formula for the lever arm between these two forces, an expression for the resisting moment against uplift of the tank is developed. Additional resistance provided by the weight of the tank shell, the bending moment distribution in the tank wall, and other sources are ignored.

The overturning moment on the tank is calculated in a manner similar to that used in the API methodology, except that the convective component of the overturning moment is neglected, in the interests of simplicity. This is felt to be appropriate since, because of phase differences between the impulsive and convective components, the convective portion is not believed to contribute much to the peak tank wall stress response, especially for more slender tanks. This omission is balanced somewhat by ignoring some portion of the overturning resistance, as described above, and by the use of a slightly conservative expression for the height of the center of mass of the fluid.

The tank is deemed to be stable if the resisting moment, M_{RES}, is greater than the overturning moment induced by the earthquake, M_{OT}. Expressed alternately, the tank is stable if the limit impulsive acceleration, C_{eq}, calculated by equating M_{OT} with M_{RES}, is greater than the earthquake-induced peak spectral acceleration at 2% of critical damping.

The method for evaluating unanchored storage tanks included herein is based on that of Manos but includes some important variations. The most notable of these are

(a) Tank anchorage is recommended in zones of high seismicity whenever the ratio of safe operating height to tank diameter exceeds two (Dowling and Summers 1993). Based on the data from Manos (1986) and the higher level of risk for taller tanks, this is believed to be the upper limit of applicability of the Manos method.

(b) The allowable compressive stress in the tank shell should not exceed 75% of the theoretical buckling stress, as presented in Manos (1986), nor should it exceed the material yield strength. This last requirement is significant for thicker-walled tanks. Note that under certain circumstances an increase in the allowable compressive stress beyond 75% of the theoretical buckling stress may be justified (Dowling and Summers 1993). Examination of Manos's experimental and observational data indicates that an increase may be justified for the types of tanks encountered at petrochemical facilities. In any event, the compressive stress should never exceed the material yield strength.

(c) The compressive force in the tank shell should not exceed the total weight of the fluid contents (Dowling and Summers 1993). This has the effect of imposing an upper bound on the resisting moment.

A final note of caution: The critical formula for term C_{eq} (the limit impulsive acceleration), as presented by Manos in Equation (9) of his paper (not shown below), should include the term (m_t/m_1) not (m_1/m_t), as originally presented, where m_1 is the liquid impulsive mass and m_t is the total liquid mass. Correction to this formula was made in the form of an erratum to the original paper (Manos 1986).

The expression for overturning moment, M_{OT}, from Manos (1986) in US customary units is given by

$$M_{OT} = 1.29\rho_w GH^2 R^2 C_{eq} \frac{m_1}{m_t} \left(\frac{H}{R}\right)^{0.15} \tag{7-1}$$

where

ρ_w = Mass density of water (=62.4/g)
G = Specific gravity of contained liquid
H = Tank liquid height
R = Tank radius
C_{eq} = Limit impulsive acceleration (in g)

and a curve for the ratio (m_1/m_t) is given in API 650 Appendix E for the ratio (W_1/W_T), where W_1 is the impulsive liquid weight and W_T is the total liquid weight. The expression for resisting moment, M_{RES}, is given by

$$M_{RES} = 0.48SERt_s^2 \left(\frac{t_s}{t_p}\right)^{0.1} \left(\frac{R}{H}\right)^{n-0.15} \tag{7-2a}$$

where

t_s = Shell thickness
t_p = Bottom plate thickness
n = 0.1 + 0.2 $(H/R) \leq 0.25$
E = Young's modulus
S = Foundation deformability coefficient

However, from Item (b) of this subsection, an upper bound on the resisting moment, assuming the compressive stress in the tank shell just equals its yield stress, F_y, is given by

$$M_{RES} = 1.06 SF_y R^2 t_s \left(\frac{t_s}{t_p}\right)^{0.1} \left(\frac{R}{H}\right)^{n-0.15} \qquad (7\text{-}2b)$$

Also, from Item (c) in this subsection, an upper bound on the resisting moment when the compressive force in the tank shell just equals the weight of the fluid contents [assuming a lever arm of $1.25\ R(H/R)^{0.15}$, as expressed in Manos (1986)], is given by

$$M_{RES} = \pi R^2 H g \rho_w G x 1.25 R \left(\frac{H}{R}\right)^{0.15} \qquad (7\text{-}2c)$$

As can be seen from the above, the resisting moment, M_{RES}, should be taken as the smallest of the expressions given in Equations (7-2a), (7-2b), or (7-2c).

An important feature of the modified Manos methodology is the use of a foundation deformability coefficient (S). This should be taken as 1.0 for tanks founded on more rigid materials, such as concrete, asphalt rings, or pads, and 1.2 for materials founded on more pliable materials, such as crushed rock, sand, wood planks, or soil. The effect of this is that the size of the compressive stress zone is larger for a soft foundation than for a rigid foundation. This enables the development of a larger limit-resisting moment (subject to the limitation imposed by the total weight of the fluid, as described above) and, consequently, enables the tank to withstand a larger seismic acceleration. This contrasts with the API methodology, where the soil type has no influence on the resisting moment, but where a softer soil leads to a larger convective acceleration and hence an increased overturning moment.

Table 7-1 shows a comparison of the results of an evaluation of a 35 ft (10.7 m) diameter, 30 ft (9.1 m) high tank, filled to a height of 26 ft 4 in. (8.0 m), using the modified Manos and API methodologies. The table shows that the API approach would require either a reduction in fill height by about 19% to 21 ft 4 in. (6.5 m) or tank anchorage, whereas the modified Manos method indicates that the seismic safe operating height can be increased to 23 ft 8 in. (7.2 m). Hence the required reduction in fill height is reduced from 5 ft (1.5 m) to 2 ft 8 in. (0.8 m). Because the latest version of API 650 uses the SRSS combination method, the difference between the two methods is not as significant as it has been under older editions of API 650.

A situation could be encountered where loss of contents of a single critical tank (or several critical tanks) is of concern. In this case, the structural integrity of the nearby tanks, containing relatively harmless materials, may only be of concern in so far as their failure could adversely impact the adjacent more critical tank(s). Although an elephant foot buckling type of failure of one of the surrounding tanks would not pose a threat to the integrity of the critical tank, a gross failure, such as overturning, could lead to an impact with the critical tank, possibly leading to rupture of the tank wall and release of its contents. In such cases, the surrounding tanks need to be evaluated to ensure that they have an adequate margin of safety against overturning or collapse (but not necessarily against buckling). For such a

Table 7-1. Comparison of API and Modified Manos Results.

<u>Typical Tank</u>

Diameter	35 ft	Product:	Vinyl Acetate
Height	30 ft	Fill Height:	26 ft 4 in.
		Roof Weight:	14.2 kip

Shell Properties: All courses: 0.18 in. thick
Bottom Plate: 0.22 in. thick

Specific Gravity:	0.93

S_s:	1.5
S_1:	0.6
Site Class:	D
Importance Factor:	1.0

<u>Site-Specific API Approach</u>	<u>Modified Manos Approach</u>
Instability Ratio: 2.09	Foundation Deformability Coefficient (*S*): 1.2
	Limiting Acceleration: 0.74 g
Therefore, unstable.	Earthquake-Induced Acceleration (S_{DS}): 1.00 g
	Therefore, unstable.
Seismic Safe Operating Height: 21 ft 4 in.	Seismic Safe Operating Height: 23 ft 8 in.

Note: Metric units not presented, because Manos (1986) only presents results for U.S. customary units.

situation, the modified Manos criteria could be relaxed still further (Dowling and Summers 1993). Examination of the data on which the Manos methodology is based (Manos 1986) suggests that failure/collapse may be less likely in cases where the ratio of overturning moment to restoring moment is less than two. Further, because local buckling of the tank wall is less of a concern, it may be appropriate to increase the allowable compressive stress in the tank shell by 33% from that presented earlier (i.e., up to the theoretical buckling stress), with the caveat to remain that the compressive stress should not exceed the material yield strength. Again, however, the compressive weight in the tank shell should not exceed the total weight of the fluid contents. These recommendations are based on limited data and should be used with care (Dowling and Summers 1993), but are supported (in principle) by research presented by Peek and El-Bkaily (1991), which suggests that the ultimate seismic overturning moment resisted by a subject 100 ft (30.5 m) diameter by 40 ft (12.2 m) high tank was 31% higher than the overturning moment at which elephant's foot buckling began.

7.4.4 Freeboard Requirements

Tanks with insufficient freeboard may have their fixed or floating roofs damaged by sloshing fluid. In fixed roof tanks, the sloshing fluid can impact and damage rafters and buckle the tank shell. Floating roofs can tilt with the sloshing wave and, if insufficient freeboard exists, the roof's seal can be damaged, or the roof may impact the access platform. To prevent this damage, sufficient freeboard may be provided to accommodate fluid sloshing. The height of the sloshing wave can be calculated using the method presented in ASCE 7 and API 650 Appendix E.

7.4.5 Uplift Calculations

When assessing the vulnerability of a tank's piping, walkway, and stairway attachments (as shown in Figure 7-6), one should consider that the tank may be subject to uplift. Using the values provided in API 650 Appendix E can be considered prudent and conservative. Actual expected values are a function of tank size, aspect ratio, fill height, and local seismicity and soil type. If explicit calculation of tank uplift is required, API 650 Appendix E provides a method.

Note that both AWWA D100 and API 650 indicate that uplift does not occur when the anchorage ratio is less or equal to 0.785. These standards use this criterion for evaluating overturning only. This criterion should not be used for assessing piping and other attachments. The minimum piping displacements provided in these standards should be used instead.

The method in Priestley et al. (1986) was proposed by a study group for the design of storage tanks. The group did an extensive review of the available literature at the time to come up with their proposed methodology. It includes provisions for rectangular and concrete tanks. Some differences between this and other methods include the following:

- Impulsive and convective components are combined by the SRSS method instead of algebraically. API 650 Appendix E and AWWA D100 also combine the impulsive and convective components by the SRSS method.

- Tank bottom uplift is based on a model that incorporates both the bending and the in-plane forces in the bottom plate.

- Method is iterative, equating the overturning moment with the resisting moment.

- Analysis is based upon ultimate loads rather than working loads.

As mentioned earlier, this method is more involved than either API 650 or the method developed by Manos. As Priestley et al. (1986) state, the calculated displacements become less accurate as the displacements increase. Good engineering judgment should be used in these cases.

7.4.6 Riveted and Bolted Tanks

Riveted and bolted tanks have the additional failure mode of tank shell splitting, which does not occur in welded tanks. Tank shell splitting is believed to occur from excessive hoop tension and poorly proportioned joints. A quick assessment of this failure mode would be to compare the cross-sectional strength of the bolts or rivets with that of the steel plate. If the bolts or rivets are stronger than the surrounding plate, the shell should behave with ductility. If additional analysis is required, the bolted/riveted section strength can be compared with the hoop tension predicted by AWWA D100 or Priestley et al. (1986). API 653 (2014) provides capacities of riveted sections.

7.4.7 Fiber-Reinforced Plastic (Fiberglass) Tanks

Fiberglass does not have the ductility typically associated with steel. Fiberglass properties are also anisotropic, that is, the strength in a direction parallel to the grain is different to that perpendicular to it. Presently, no standard exists for designing or assessing fiberglass tanks for seismic loadings. ASME BPVC (2015) and ASTM D3299 (2010) provide guidance on nonseismic design. Many variables are involved in determining the strength and ductility of the fiberglass composite material, including the properties of the resin matrix and the angle of winding of the fibers. Seismic loads should be determined from API 650, with due consideration given to expected tank performance and ductility, and allowables obtained from manufacturers' recommendations or test results.

7.5 MITIGATION OF SEISMIC EFFECTS

Mitigating a seismic hazard can be quite involved or relatively simple. For overconstrained piping, additional bends or a flexible section of piping may be added. Stairs and walkways can be solely supported by the tank shell. However, tank wall stability can be more difficult to correct. No one method will work all the time and operating and construction economics should be considered.

Where the tank is found to be structurally inadequate (as determined by exceedance of the modified Manos criteria or gross exceedance of the API criteria), any of the following retrofits may be implemented:

(a) Reduce the fill height: This is the simplest and most commonly recommended retrofit and should also be considered in cases where the available freeboard is found to be inadequate. Note that reduction of fill height can have a significant effect on the economics of tank storage.

(b) Increase the shell thickness and/or the bottom plate or annular ring thickness.

(c) Anchor the tank in accordance with the provisions specified in API 650.

(d) In lieu of anchorage, prevent uplift of the tank by stiffening the tank base through the installation of a concrete slab within the tank shell, or by other methods. This method is relatively untried but may have the same effect as anchoring the tank.

7.6 CONSIDERATIONS FOR FUTURE INVESTIGATION

This committee did not review all aspects of seismic tank performance. Two areas that would warrant further evaluation include

(a) Calibration of the method presented in API 650 to determine uplift of tanks with actual performance data and/or detailed analytical results.

(b) Development of seismic design and evaluation procedures for fiberglass tanks.

7.7 DESIGN OF NEW TANKS

Seismic design of new tanks is covered in API 650 Appendix E, which, being conservative, is a good candidate for design. The following modifications or additions are proposed to address other shortcomings.

Sloshing Fluid

Earthquakes cause the upper portion of the contained fluid to slosh. The height of the sloshing fluid can be calculated by an equation in ASCE 7 and API 650 Appendix E. This height should be used for freeboard as required in ASCE 7 and API 650 Appendix E to prevent or minimize floating roof damage or spillage of product in floating roof or open roof tanks. The sloshing fluid can also impact and damage rafters and supporting columns of fixed roofs. Wozniak and Mitchell (1978) show how to calculate this wave force on columns. Malhotra (2005) presents a method to calculate this wave force on the fixed roof. The impulsive force on the tank is also increased when the sloshing liquid is confined by the fixed roof due to inadequate freeboard. ASCE 7 requires this additional impulsive force to be considered in the design. Malhotra (2005) shows how to calculate this increased impulsive force.

Overturning Moment on the Foundation

The hydrodynamic forces that create the overturning moment on the tank walls also act on the tank bottom and hence on the foundation. This additional overturning moment (ASCE 1984) should be included when designing the structural portion of the foundation and piles. API 650 presents guidance on calculation of the "slab" moment.

Tank Movement

Tank uplift during earthquakes can damage attached piping and other appurtenances. The same provisions discussed in Section 7.4.5 may be used in design. Additionally, anchored tank appurtenances may be designed for some level of anchor bolt stretch. A (working stress) displacement value of 1 in. (25 mm) is required in ASCE 7 and API 650 Appendix E.

API 650 states that piping attached to the tank bottom that is not free to move vertically shall be placed a radial distance from the shell/bottom connection of 12 in. (300 mm) greater than the uplift length predicted by the API 650 uplift model. The API 650 uplift model, however, may underpredict the amount of radial uplift (Manos 1986; Dowling and Summers 1993). Changing this requirement to that predicted by Priestley et al. (1986), twice the API 650 model, or an interaction of allowable vertical movement with the distance placed radially from the shell, may be prudent.

Walkways between tanks should be designed to accommodate relative movement of the tanks. API 650 requires that the calculated movement be amplified by a factor of 3.0 and added to the amplified movement of the adjacent structure. In lieu of a more rigorous analysis, a walkway should be designed to accommodate a total of 12 to 18 in. (300 to 450 mm) of movement, at least in zones of high seismicity. This is based upon limited experience data from past earthquakes and is thought to be conservative for tanks with small height-to-diameter ratios. For anchored tanks, this movement should be reduced further.

Anchored Tanks

API 650 or other methods may be used for design. Attached ringwalls should be designed appropriately. Anchoring a tank to a small ringwall and not developing the forces into the soil by the weight of the ringwall or with piles should be viewed with caution. Anchor bolts need to be designed such that they behave in a ductile manner, both in terms of the force transfer to the shell and pullout from the concrete foundation. API 650 presents some guidance in this regard.

References

API (American Petroleum Institute). 2014a. *Tank inspection, repair, alteration, and reconstruction*, 5th ed. Washington, DC: API.

API. 2014b. *Welded steel tanks for oil storage*, 12th ed. Washington, DC: API.

ASCE. 1984. *Guidelines for the seismic design of oil and gas pipeline systems*. New York: ASCE.

ASCE. 2016. *Minimum design loads and associated criteria for buildings and other structures*. ASCE/SEI 7-16. Reston, VA: ASCE.

ASME. 2015. *Boiler and pressure vessel code*. New York: ASME.

ASTM. 2010. *Standard specification for filament-wound glass-fiber-reinforced thermoset resin chemical-resistant tanks*. ASTM D3299-10. West Conshohocken, PA: ASTM.

AWWA (American Water Works Association). 2011. *Welded steel tanks for water storage*. AWWA D100. Denver, CO: AWWA.

Dowling, M. J., and P. B. Summers. 1993. "Assessment and mitigation of seismic hazards for unanchored liquid storage tanks." In *Proc., Independent Liquid Terminals Association Bulk Liquid Transfer and Above-Ground Storage Tank Terminals Conf.*, Houston.

EERI (Earthquake Engineering Research Institute). 1990. *Loma Prieta reconnaissance book*. Oakland, CA: EERI.

EERL (Earthquake Engineering Research Laboratory). 1971. *Engineering features of the San Fernando Earthquake, February 9, 1971*. Pasadena, CA: EERL.

EQE. 1986. *Effects of the March 3, 1985 Chile Earthquake on power and industrial facilities.* Oakland, CA: EQE International.

Housner, G. W. 1957. "Dynamic pressures on accelerated fluid containers." *Bull. Seismol. Soc. Am.* **47** (1): 15–37.

Housner, G. W., and M. A. Haroun. 1980. "Seismic design of liquid storage tanks." In *Proc., ASCE Convention and Exposition*, Portland, OR.

Malhotra, P. K. 2005. "Sloshing loads in liquid-storage tanks with insufficient freeboard." *Earthquake Spectra* **21** (4): 1185–1192.

Manos, G. W. 1986. "Earthquake tank-wall stability of unanchored tanks." *J. Struct. Eng.* **112** (8): 1863–1880. (1986)**112**:8(1863).

National Research Council, Committee of the Alaska Earthquake of the Division of Earth Sciences. 1973. *The great Alaska earthquake of 1964.* Washington, DC: National Research Council, National Academy of Sciences.

NZS (New Zealand Standard). 1986. *Code of practice for concrete structures for the storage of liquids.* NZS 3106. Wellington, NZ: Standards Association of New Zealand.

Peek, R., and M. El-Bkaily. 1991. "Post-buckling behavior of unanchored steel tanks under lateral loading." *J. Pressure Vessel Technol.* **113** (3): 423–428.

Priestley, M. J. N. (ed.), B. J. Davidson, G. D. Honey, D. C. Hopkins, R. J. Martin, G. Ramsay, et al., 1986. *Seismic design of storage tanks.* Wellington, NZ: Study Group of the New Zealand National Society for Earthquake Engineering.

Summers, P. B., and P. A. Hults. 1994. "Seismic evaluation of existing tanks at grade." In *Proc., American Power Conf.*, Chicago.

Wozniak, R. S., and W. W. Mitchell. 1978. "Basis of seismic design provisions for welded steel oil storage tanks." In *Proc., 43rd Midyear Meeting on Session on Advances in Storage Tank Design, API Refining*, Toronto.

CHAPTER 8

Earthquake Contingency Planning

8.1 INTRODUCTION

Although not explicitly a part of petrochemical and other industrial facility design, a section on contingency planning was included due to the importance of the topic. This section does not give specific guidance on how to author contingency plans, but merely outlines some general points that in-place contingency plans should address.

8.2 PURPOSE

Every facility located in a seismic region is expected to have a contingency plan in place. All parties involved in the response to a seismic event must be familiar with the response plan to utilize it to its fullest potential. This section offers suggestions that such a plan should include.

The primary purpose for assembling an earthquake emergency contingency plan is to have a rapid, rational, and structured response to a seismic event. A contingency plan will allow the following to be done efficiently and safely:

(a) Emergency response to maintain public, personnel, and plant safety;

(b) Organized inspection of facilities for earthquake structural damage and conveyance of results to management to minimize business interruption; and

(c) Focusing of critical resources for recovery effort.

8.3 SCOPE OF RESPONSE PLAN

This Guideline aims to cover the organization of personnel and resources to perform structural damage assessment and damage control after a seismic event. An effective response plan should cover all of the following elements:

- Pre-earthquake preparation,
- Event recognition,
- Command and control system / mobilization of inspection team,
- Roles and responsibilities of team personnel,
- Inspection methodology, and
- Assembling of inspection data / results reporting.

This section concentrates mainly on Items a through d, with e and f covered in more detail in Chapter 9.

8.4 PRE-EARTHQUAKE PREPARATION

(a) The key to effective response to any disaster situation is organization and planning before the event actually occurs. As likely scenarios could entail loss of water, communications, and firefighting ability; inability to reach the site; release of hazardous materials, etc., taking a proactive role in planning for any emergency is imperative. Some of the most critical areas to address arepre-event assessment of seismic risk

A key component in planning response to an earthquake emergency is to understand the type of damage to expect and where it may occur. Performing a seismic assessment of a facility is an ideal way to assess the inherent risk at the facility. Another benefit of performing a seismic assessment is allowing structures to be prioritized according to risk. This risk hierarchy can then be incorporated into the response plan, allowing the highest-risk structures to be inspected first. This risk hierarchy list needs to be a "living" document and must have an owner to maintain it.

(b) Formal response plan

Another key component of effective response to a disaster is a written response plan that outlines

- Organization of inspection teams,
- Command structure of teams,
- Listing of key contacts, and
- Risk-prioritized listing of structures.

The response plan must effectively outline the team's organization, command structure, and role in the assessment process to be useful. Its contents must be effectively communicated to and understood by personnel who will be involved in the assessment effort. Drills are an excellent method to help implement the intents of the document.

Setting up standing agreements with support resources such as local engineering firms, local emergency response agencies, and professional organizations such as ASCE is often useful. Such standing assistance agreements can help streamline the response process.

(c) Pre-earthquake training

For inspection personnel to perform effectively, they must be trained in post-earthquake damage assessment. All inspection personnel should undergo ATC-20 (1988) or similar training for building assessment and similar training geared toward inspection of industrial structures.

(d) Inspection materials

For quick response, inspection materials (i.e., paper, pencils, clipboards, cameras, flashlights, batteries, etc.) and structural priority information and posting materials should be kept in a secure location that is to be accessed only in the event of an earthquake. The post-event scenario is inherently confusing without having to find inspection materials. Including search and rescue equipment may be desirable. Water and food supplies should also be kept in the secured location, as shelter-in-place conditions may occur.

8.5 INCIDENT RECOGNITION

Before beginning any response or assessment effort, some formal recognition of the event must take place. Recognition of the event will allow response plans to be put into effect. Creating a matrix or classification system to describe the extent of the event (i.e., low, moderate, severe) is advisable. This matrix could be based on the extent of the observed damage. By doing this, each level of alert can then be paired with an appropriate level of response. The obvious benefit is to employ the appropriate level of response for the size of event that has occurred. For additional information, refer to Chapter 9.

8.6 COMMAND AND CONTROL/MOBILIZATION SYSTEM

A clear command and control structure must be maintained to respond to an emergency effectively. Key contacts and decision makers must be clearly identified, and a hierarchy of authority must be clearly understood. The response plan should specifically designate key contacts and decision makers and backups for those individuals in case the primary contacts are unavailable. It should also describe how and to what decision makers inspection results should be communicated. Finally, the emergency operations center(EOC) should be located in a safe area, preferably on the perimeter of the site.

After an earthquake, short-term (rapid assessment) and longer-term (detailed assessment) responses will be necessary. Immediately following an event, rapid assessment teams will need to deploy to identify obvious or immediate problems. These problems might include failures, environmental releases, fires, etc. Concurrently, the incident commander will need to start mobilizing structural expertise to (a) advise plant personnel on the course of action to stabilize immediate failures or crises; and (b) begin performing a detailed inspection.

Once the rapid assessment teams have reported to the incident commander and the seriousness of the incident has been determined, the appropriate level of response must be mobilized. The incident commander must have trained personnel available to make decisions regarding short term stabilization of failures or crises. Examples may include advising on temporary bracing/shoring, coordination of multidisciplinary problems, advising on plant shutdown, etc.

The response plan must contain instructions for the team members once the level of response needed has been determined. These instructions must include where and when to report. Additionally, the plan must include details instructing team members on whom to report to once they reach the facility. Some contingencies need to be addressed to account for team members' inability to reach the site, difficulty in communications, etc., or any foreseeable difficulties that may be encountered.

8.7 ROLES AND RESPONSIBILITIES OF TEAM PERSONNEL

After the inspection team has been assembled, they must know what is expected of them and what responsibilities they will be held accountable for. The response plan should include a model team structure to show each team member's role. The desired qualifications for each role in the inspection effort should be outlined in the response plan, while recognizing that circumstances may not allow a perfect matching of skills to positions. Having an operating representative as part of the inspection team is advisable. The plan should also clearly outline lines of communication and authority within the team structure. See Chapter 9 for additional information.

8.8 INSPECTION METHODOLOGY

The response plan should address the methodology to employ for damage assessment. Things to specify include

- Inspection criteria to use (ATC-20 / other),
- Posting criteria for buildings / structures,
- Authority of postings,
- Dealing with aftershocks, and

- Repair authority.

Chapter 9 addresses this item in more detail.

8.9 ASSEMBLING OF INSPECTION DATA/REPORTING RESULTS

After performing field inspections, posting buildings, and making structural evaluations, data and results must be assembled to report it to the incident commander. Having predesigned forms to help standardize the data collected and make it easier to assemble after the inspections is desirable. These forms should be stored in a secured cabinet, such as the one discussed in Section 8.4. It also allows documentation of damage to be kept for future review and/or analysis. Using standard forms makes the task of assembling and transmitting the information to the appropriate decision maker much more efficient. Also, these forms should include some method of incorporating secondary inspections after the inevitable large aftershocks. Chapter 9 provides additional information.

Reference

ATC (Applied Technology Council). 1988. *Procedures for postearthquake safety evaluation of buildings.* ATC-20. Redwood City, CA: ATC.

CHAPTER 9

Post-Earthquake Damage Assessment

9.1 INTRODUCTION

This chapter provides guidance on post-earthquake damage assessment of facilities considered to be in need of structural investigation after an earthquake. After every significant seismic event, an immediate and complete walkdown of the petrochemical or other industrial facility in question is the optimum response. However, although such action is strongly advised, this may be impractical due to cost and manpower constraints. For these reasons, two factors must be considered with regards to the scope of and necessity for post-earthquake damage assessments: prioritization of items to be assessed after an earthquake and determination as to what "size" of event warrants an assessment.

9.1.1 Assessment Priorities

The major safety/operational concerns of operators of petrochemical and other industrial facilities fall under the following three priorities:

First Priority: Health and human safety.

Second Priority: Damage to the environment.

Third Priority: Other items with significant economic impact.

Under the first-priority item, the major risks to health and human safety associated with earthquakes (in no particular order) are the following:

(a) Fires,

(b) Explosions,

(c) Collapsed and damaged structures, and

(d) Release of hazardous materials.

Under the second-priority item, the major risk to the environment associated with earthquakes is the release of environmentally sensitive products. Such releases are generally related to the following occurrences:

(a) Storage tank failure,

(b) Pipeline rupture or failure, and

(c) Failure or collapse of other vessels containing environmentally sensitive products.

Note that failure in the above discussion also refers to significant leakage of tanks, pipelines, and vessels, and is not limited to complete collapse.

Under the third-priority item, the major additional cause of significant economic impact to a facility operator is a loss of production capability. Occurrences that contribute to loss of production capability (in addition to those listed in the first two priority items) are the following:

(a) Loss of power,

(b) Loss of communications, and

(c) In-place emergency shutdown procedures activated by the earthquake.

Auxiliary structures that also require investigation following an earthquake include the following:

(a) Utility systems that would be required to operate following an earthquake to maintain the facility in a safe condition, for example, firewater and emergency power systems;

(b) Adjacent systems, structures, or components whose structural failure or displacement could result in the failure of systems that pose a risk to health and human safety, contain hazardous materials or environmentally sensitive products, or are particularly valuable from an operations point of view; and

(c) Any structures, systems, or components that, from an operator viewpoint, are required to remain functional during and after a major earthquake.

9.1.2 Assessment Triggers

No absolute quantitative measure of earthquake severity can be given as a guideline as to when a post-earthquake damage assessment should be performed. Much depends on whether any significant damage has been reported at the facility and whether the earthquake event was serious or damaging. Ideally, a post-earthquake damage assessment should take place after a seismic event of any significant magnitude. The following examples (either occurring at the facility or in the immediate vicinity of the facility) should trigger the need for a post-earthquake response:

• Injury to personnel;

• Structural collapse;

• Release of hazardous materials or environmentally sensitive products;

• Fire;

• Explosion;

- Automatic system shutdown; and
- Loss of power, utilities, or communications.

Any of these occurrences may indicate more serious problems, and a post-earthquake damage assessment should be initiated immediately. In addition to the structural investigation, mechanical and systems assessments may also be required.

9.2 PRE-INVESTIGATION ACTIVITIES

The following safety-related actions, which should occur immediately after a significant seismic event, should precede a post-earthquake damage assessment:

- Accounting for the safety and welfare of all persons on the site (employees, contractors, and visitors);
- Controlling all immediate dangers, such as fires, collapsed structures, and so forth; and
- Determining safety for nonemergency response trained personnel, that is, the engineers performing the post-earthquake damage assessment, to enter the facility.

Preplanning inspection activities is essential and should include a meeting with plant personnel familiar with the facilities under review. To help ensure that all pertinent topics are addressed, a list of questions should be prepared before this meeting. One individual should be appointed to write notes for a record of conversation, or a recording device may be used. Topics to be discussed with plant personnel include the following:

- Plant operating history before and after the earthquake,
- Corrective action taken by the operators during or after the earthquake,
- Maintenance or repairs that may have been made before the inspection,
- Any problems experienced with operability and functionality,
- Any equipment failures during or after the earthquake,
- Equipment accessibility,
- Hazardous areas,
- Safety precautions, and
- Existence of any instrumentation on site that recorded the earthquake motion.

Corrective actions, maintenance, and other repairs made at the facility due to seismic activity should receive particular attention. The adequacy of such repairs should be checked, and the location of these corrective activities may indicate additional areas of concern.

During preinvestigation activities, the inspection team should also discuss and determine what areas of the plant, equipment, or components will be inspected and in what order. Plant arrangement drawings should be obtained and marked with the location of specific components selected for evaluation. If more than one inspection team is involved, areas of responsibility should be divided as needed.

A prudent reminder for all those involved with the inspection is that the plant has experienced a seismic event. Before beginning inspections, plant personnel should visually inspect the plant for any existing hazardous condition (e.g., areas contaminated by chemical spills). If any hazardous condition is found, this area should be noted and inspection team members advised accordingly. Inspection of these areas should be performed only if precautions to protect the safety of the inspection team members can be and have been taken.

One of the most important functions of the pre-investigation phase is to thoroughly question operators as to the extent of the damage. Usually, the operators will have already performed a quick inspection of the facility and will be able to provide exact locations of known damage. Such areas should be the first priority of the inspection team. If a seismic hazard mitigation program has already been undertaken at the facility, then the structures known to be seismically vulnerable and their weak links should be the second priority of inspection. If no such program has been previously undertaken, then the vulnerable structures will not be known a priori, and the assessment team should sweep the entire facility systematically, while concentrating on the assessment priorities as discussed previously in Section 9.1.1. Attention should be given to types of items that have performed poorly during past earthquakes, such as older structures. In any event, plant operators will be the key to prioritizing the assessment.

9.3 PERFORMING FIELD INSPECTIONS

Inspections of the facility should be documented with field notes and accompanying photographs. Checklist-style forms may be used; see Chapter 6 for examples. Although inspection team members may prefer taking notes to filling out checklists, using these forms can provide a greater measure of consistency.

Measurements should be made with a tape measure as much as is practical. For inaccessible components, a visual estimate of dimensions should be made. When visual dimensions are used, the documentation provided should reflect this fact. For some items, an as-found sketch may be needed to identify all of the parameters investigated. Complete information, as required, should be provided on the sketch regarding dimensions, sizes, component types, etc. At least one photograph of each structure or component investigated should be taken. A labeling system should be implemented to match the photographs to the inspection notes. If permissible, a video camera can be used. The video camera will generally provide a better visual aid for historical data, and the voice recording unit of the camera can be used to narrate condition and type of equipment being inspected.

Above all, any damage resulting from the seismic event should be described in detail. Damage should be assessed immediately at each individual location, and some engineering basis to explain the failure should be provided and noted, if possible. The inspection team should try to come to a consensus as to any necessary remedial measures.

9.4 EVALUATION OF LOAD-CARRYING SYSTEMS

Of paramount importance when performing a post-earthquake damage assessment is the ability to determine what exactly constitutes a given structure's gravity and lateral force resisting systems. A structure whose gravity system has been damaged, and its ability to continue to carry vertical (dead) loads is in question, is a very hazardous condition since collapse may be imminent. This becomes even more critical when the likelihood of aftershocks is included. Such a structure should be immediately flagged, and suitable measures taken, such as barricades, evacuation, immediate stabilization, demolition, etc.

Structures with intact gravity systems, but substantial damage to their lateral force–resisting system, for example, buckled or damaged braces, should also be regarded as potential hazards, due to the possibility of future aftershocks. Although such structures are generally not as significant a hazard as those with damaged gravity systems, immediate steps should be taken to restore the integrity of the original lateral force–resisting system.

Deciding whether a structure has a damaged vertical and/or lateral force–resisting system often rests on the question regarding the structure's ability to withstand an aftershock of equal or greater magnitude. If an experienced damage assessment team does not believe the structure is capable of surviving such an event, immediate steps should be taken, and entering the structure or its immediate vicinity should only be done on a limited as-needed basis. Continued operation should be justified on a case-by-case basis. Some items of particular concern include the following:

- Loss of integrity of vertical and/or lateral force–resisting systems;
- Steel structures: yielding, hinging (top and bottom of columns), web buckling and crippling, brace buckling, weld cracks, and so forth;
- Concrete structures: excessive cracking, spalling, exposed rebar, damaged rebar, hinging, crushing, etc.
- Anchors and foundations: anchors and/or bolts pulled out of footings, stretched anchors, large depressions around column bases, etc.

- Equipment: visible punctures, leaks, cracks, toppled items, loose fittings and valves, contents under high pressure, weld cracks, enclosed areas where build-up of explosive gases may occur (confined spaces), etc.

- Roadways: clear of large debris, downed power lines, "continuous surface" (no faulting), etc., and

- Evidence of liquefaction, fault rupture, significant settlement, landslides, and tsunami/seiche.

Specific to stretched anchors described in Item d, Soules et al. (2016) report that the CAP Acero steel mill in Chile considers a maximum stretch of 5% for anchor rods to be acceptable from the standpoint of energy dissipation. The 5% limit is based on testing (Powell and Bryant 1983).

9.5 IDENTIFICATION OF DAMAGED STRUCTURES

A facility-specific "tagging" system should be considered to identify systems, structures, or components judged to be vulnerable by the inspection team. A system developed for post-earthquake evaluation of buildings and used relatively successfully following major earthquakes is described in ATC-20 (1988). That system essentially involves placing green (safe), yellow (enter with caution for restricted use), and red (do not enter) tags on inspected buildings. Such a system may require modification for use in a petrochemical or other industrial facility setting, due to the congestion and (often times) indistinguishability of different structures, and it should be understood that multiple tags or tape may be required to rope off those sections deemed unsafe by the assessment team.

9.6 DOCUMENTATION

The objective of the inspection team's effort at each facility is to collect all information needed to document equipment and structural performance during the earthquake. Figure 9-1 shows, in the form of a recommended table of contents, the organization and content of a report that can be produced for each facility visited (EQE 1986). Note that the contents of this report can be separated into those sections that should be substantially completed during the field inspection and those that can be completed later.

The final report should reference published information on the earthquake severity in discussions of the ground motion at each facility, summarize information of the inspected equipment and structures at each facility, discuss generally the performance of each type of inspected equipment, and describe in detail any damage that was noted during the inspections, along with any required or recommended retrofits.

TABLE OF CONTENTS

1. INTRODUCTION AND PURPOSE. Generic description of purpose of report, which is to document post-earthquake inspection of facility after earthquake event. This section can be completed after inspection.

2. FACILITY DESCRIPTION. A brief overview of facility, including function. Description of major structures, age, exact location, operating history, etc. During the plant visit, collect information necessary to include in this section: plant literature, drawings, tape recordings of interview with personnel, etc.

3. GROUND MOTION AT FACILITY. Unless records during the plant visit can be obtained, expect to write this section later using published reports of records throughout the area (e.g., those published by USGS).

4. OPERATIONAL IMPACT OF EARTHQUAKE ON FACILITY. Interviews with plant personnel will form the basis for this section. If possible, obtain copies of operator logs during the plant visit.

5. EARTHQUAKE PERFORMANCE. This section will form the main body of the report and should be completed during the field inspection. This section should include recommendations for remediation, if required. Responsibility for originating various subsections should be divided up among walkdown team members.

6. CONCLUSIONS/RECOMMENDATIONS. Briefly summarize the primary findings of the team, including recommended actions. This section may be written later.

7. APPENDIX A. The appendix should contain photographs, plans, drawings, etc., which are referenced within the body of the report.

Figure 9-1. Example report table of contents.
Source: EQE (1986).

9.7 INSPECTION TEAM

Each inspection group should consist of a minimum or two (2) individuals with the following qualifications:

(a) Petrochemical or other industrial facility walkthrough inspection experience; and

(b) Experience in seismic design and analysis of equipment, systems, and structures found in petrochemical or other industrial facilities.

Inspection team members should have undergone training in post-earthquake damage assessment and in the procedures to document such assessments at the facility in question.

The structural portion of a post-earthquake damage assessment should be conducted by experienced engineers familiar with seismic design and the effects of earthquakes. They should be able to readily identify areas of weakness and prioritize remedial measures to correct severe damage and to ensure continued safe operation of the facility.

9.8 RECOMMENDED EQUIPMENT

The following items are recommended while performing post-earthquake damage assessments:

- Flashlight,
- Tape measure or folding rule,
- Compass/GPS device,
- Clipboard,
- Hard hat,
- Safety glasses,
- Hearing protection,
- Safety mask (particularly after a major seismic event and/or in areas of known hazardous material or environmentally sensitive product release),
- Construction boots,
- Plumb bob or level,
- Photo camera,
- Long-sleeve shirt or any protective clothing as mandated by site personnel,
- Plot plans and facility road map,
- Handheld video camera (optional), and
- Tape recorder (optional).

Note: Check camera, flash, video, and tape recorder use with facility operators prior to their use at the facility. This should be done not only for confidentiality purposes but also because some camera equipment is considered an ignition source in certain hazardous areas or could trigger certain fire detection systems.

References

ATC (Applied Technology Council). 1988. *Procedures for postearthquake safety evaluation of buildings.* ATC-20. Redwood City, CA: ATC.

EQE. 1986. *Post-earthquake investigation: Planning and field guide: NP-4611-SR.* Oakland, CA: EQE International.

Powell, S. J., and A. H. Bryant. 1983. "Ductile anchor bolts for tall chimneys." *J. Struct. Eng.* **109** (9): 2148–2158.

Soules, J. G., R. E. Bachman, and J. F. Silva. 2016. *Chile earthquake of 2010—Assessment of industrial facilities around Concepción.* Reston, VA: ASCE.

CHAPTER 10

Retrofit Design

10.1 INTRODUCTION

The guidelines presented in this chapter are intended to assist a design professional performing pre-earthquake seismic upgrading and modification of existing structures. These guidelines are not intended to perform as a code document; rather, they aim only to give guidance to the engineer faced with a task of retrofitting seismically deficient structures. The guidelines cover both plant structures (i.e., nonbuilding) and buildings located within the plant facility. Examples of plant structures are skirt-supported vertical vessels, horizontal vessel-supporting structures, steel frame structures supporting elevated equipment, concrete table-top frames, and so forth.

10.2 UPGRADE SITUATIONS

Generally, structures need seismic upgrading and modification for one or more of the following reasons:

(a) Existing plant or building structures have been determined to be potentially hazardous to life safety after a walkthrough and an initial screening evaluation process (see Chapter 6). The initial evaluation procedure is a cursory review of existing seismic-resisting framing structures, systems, and equipment, to determine if they are categorized as "safe," "unsafe," or "need further study." For structures and systems that are classified as safe, no further action is required. However, for those that are classified as unsafe or need further study, a more rigorous evaluation should be performed.

(b) Additions to and modifications of existing structures often trigger the need to upgrade structures. Upgrading may be to the code of record if approved by the local building official. The upgrading may be required to meet current code. However, the entire structure may not need to be checked against the current seismic code; for example, if the following conditions exist:

1. Addition is designed to the current code and is independent of the existing structure.

2. Addition does not increase seismic forces, stresses, and displacements in the entire modified structure by more than a set percentage of the design values from the code of record. Values of 5% to 25% have been used by local building officials. Industry standards recommend a specific limit; for example, ASCE 7-16 and IBC (2018) recommend a value of 10%; IEBC (2018) uses 75% of that required by the code.

3. Deteriorated conditions do not reduce the existing structure's capacity by more than 10%.

4. Addition does not decrease the seismic resistance of any of the existing structure's structural elements unless the reduced resistance is equal to or greater than that required for new structures.

5. Connections between the new and existing components meet current code requirements.

(c) Owners often upgrade existing structures to reduce the risk of business interruption losses, or when the function of the structure is changing, for example from a warehouse to a building housing personnel.

10.3 CRITERIA FOR VOLUNTARY SEISMIC UPGRADING

The intent of voluntary seismic upgrading and modification of existing structures is not to bring them up to the current codes. Rather, the goal is to provide reasonable assurance to the public, owners, and state building officials that structures and systems will survive a major earthquake without catastrophic failure, though they may suffer extensive damage. Owners may desire to retrofit their plant and building structures to a higher performance level in an effort to minimize business interruption losses. Additionally, there are situations where regulatory agencies set the retrofit standard. See Chapter 2 for additional background.

10.4 SEISMIC RETROFIT CONSIDERATIONS FOR PLANT STRUCTURES AND BUILDINGS

For both plant and building structures, it is recommended that the design professional obtain an up-to-date geotechnical report, review local seismicity, review original structural drawings, calculations, verify the vertical and lateral seismic force resisting systems, and identify the potential weak links and structural deficiencies in the structural systems.

Consideration should also be given to the following prior to developing a retrofit scheme:

- Site-specific seismic hazards including any presently known site-specific geologic hazards.
- Results of prior seismic assessment and level of retrofit significance/complexity.
- Structural use (i.e., supporting vessels containing hazardous material, or buildings that require immediate use after a major earthquake).
- "As-is" structural conditions (i.e., corrosion, missing components, altered conditions, etc.)
- Economic impacts (i.e., cost effective, retrofits could only be constructed during plant shutdown or when the building is not occupied, etc.)
- A site visit to identify any potential construction problems or interferences.
- Local jurisdictional requirements.

During the evaluation and design of the retrofit, the following additional considerations are recommended:

- Use actual dead and live loads in the analysis, not design basis loads.
- Use in-situ member strengths for calculating the member capacities. Non-destructive testing may be required to obtain the as-built strength of structural members.
- Check the existing seismic capacities (C) of structural members, connections, and equipment supports versus the seismic demand (D). If capacity is greater than demand, no further action is required. Otherwise, seismic upgrading and modification should be initiated. It is recommended that strength design methods be used to calculate capacities.
- When deformations and drift limits need to be checked against the allowable, they should be consistent with those discussed in Chapter 5.
- Alternate procedures using rational analyses based on well established analysis principles may be used. Methods such as nonlinear inelastic time history analyses would be acceptable, as long as it meets local jurisdictional requirements.

Another important consideration when performing retrofit design is the concept of proportioning within the structure. Overstrengthening areas of a structure that are currently deficient in strength can force the weak link(s) to other elements that are perhaps even weaker or more brittle. If this occurs, the impact upon the overall structural performance is often undesirable. Therefore, care must be taken so that a currently weak, but ductile, structure is not modified in a way that it becomes a structure with a brittle failure mode. In a similar manner, it is important not to create strength or stiffness irregularities either vertically or in plan.

Finally, there are some additional issues that must be considered when retrofitting a structure. Issues such as access, functionality, detailing, and constructability must be addressed. In some cases, access requirements may be the determining factor in selecting a retrofit scheme. A typical example is the use of chevron braces as opposed to cross bracing, or the use of moment frame instead of shear walls. In addition, the retrofit must be functional and not create any additional problems. An example is that of air-cooled heat exchangers, or finfan structures, atop elevated pipeways. The supports for the exchangers are often not designed for seismic loads and may lack proper detailing. When retrofitting, the structure should not be stiffened such that its new structural period approaches that of the fan motors because this may cause unintended resonance. Detailing is also a key consideration in retrofit design. One of the primary goals of retrofitting is to provide a ductile system. An example is retrofit of anchor bolts on tall process columns. Retrofit design should ensure that the mode of failure will be a ductile, (i.e., yielding of the anchor bolts, and not brittle, \or pullout from the concrete or buckling of the support skirt). Finally, constructability of the retrofits is a key consideration. Knowledge of whether bolting or welding is preferred, structure accessibility, potential interferences, etc., all will have bearing on the selected retrofit option.

10.4.1 Review of Seismic Retrofit Scheme

Once a retrofit scheme is developed, the design professional is recommended to review the scheme with the facility / plant owner and building official to determine any restrictions or issues associated with the retrofit and get their approval prior to any detailed engineering design work is performed.

For plant structures or buildings, some of the restrictions or issues may be

- Need to do retrofit while plant is in operating mode or the building is occupied,
- Restricted environment (e.g., fire hazard, dust sensitive, etc., may not be suitable for welding activities),
- Contaminated site (e.g., contaminated soil removal or lead abatement),
- Change in use or occupancy,
- Constructability and accessibility, and
- Economic impact.

A rough cost study (order of magnitude) is recommended with consideration of all known restrictions or issues to identify the most beneficial retrofit solution.

10.4.2 Seismic Retrofit Objectives

When retrofitting a structure, the main objective is to provide reasonable assurance that the retrofitted structure will perform in an acceptable manner when subject to strong ground motion. Note that the intent of retrofitting is not to bring the structure up to the letter of current code. Owners may identify structures that are critical to plant operation and retrofit them to a higher performance standard to minimize business interruption.

For buildings, ASCE 41 (2013) recommends four levels of upgrading performance: Collapse Prevention Level, Life Safety Level, Immediate Occupancy Level, and Operational Level, with the associated damage levels ranging from severe to very light.

10.4.3 Retrofit Design Requirements and Methods

Chapter 5 provides criteria to be used in the evaluation and design of seismic retrofits of existing structures. The following documents provide additional guidance for evaluating and upgrading existing structures:

- CalARP (2013);
- IEBC;
- ASCE 7, Appendix 11B;
- ASCE 41; and
- IBC.

To ensure that proper corrective actions are implemented in seismic upgrade programs, engineers should be fully aware of the latest developments in earthquake engineering. Recommended resources include but are not limited to the Applied Technology Council, the American Concrete Institute, the American Institute of Steel Construction, ASCE, FEMA, and the Earthquake Engineering Research Institute.

References

ASCE. 2013. *Seismic evaluation and retrofit of existing buildings.* ASCE 41. Reston, VA: ASCE.

ASCE. 2016. *Minimum design loads and associated criteria for buildings and other structures.* ASCE/SEI 7-16. Reston, VA: ASCE.

CalARP. 2013. *Guidance for California accidental release prevention (CalARP) program seismic assessments.* Sacramento, CA: CalARP Program Seismic Guidance Committee.

IBC (International Building Code). 2018. *International building code.* Country Club Hills, IL: International Code Council.

IEBC (International Existing Building Code). 2018. *International existing building code.* Country Club Hills, IL: International Code Council.

CHAPTER 11

New and Existing Marine oil Terminals

11.1 INTRODUCTION

This chapter describes the seismic design and analysis of marine oil terminal (MOT) wharves and piers. These structures are used to moor tank vessels and barges and to transfer liquid bulk petroleum products. This chapter addresses only pile-supported structures. It does not address sheet pile structures or the design or analysis of offshore multipoint or single-point mooring systems.

The approach described in this chapter derives primarily from the California Code of Regulations, Title 24, Part 2, Chapter 31F, otherwise known as the Marine Oil Terminal Engineering and Maintenance Standards (MOTEMS) (2016), which became state law in California in 2006 and specifically addresses marine oil terminals. MOTEMS has been used extensively for the seismic evaluation of every marine oil terminal in California and for the design of several new and replacement marine oil terminal structures. It has also been used as a criteria document on other projects, especially for seismic design.

In 2014, ASCE created the first edition of ASCE 61, which addresses the seismic design of industrial piers and wharves without public access. As part of the California Building Code, MOTEMS is updated regularly and has adopted many of ASCE 61 provisions.

The approach used in these two documents differs significantly from that of ASCE 7-16. ASCE 7 provides minimum load criteria for design strength and allowable stress limits. This is not the same as is commonly used in the port/harbor industry. Compared with multistory buildings or multispan bridges, wharves and piers are usually rather simple structures. However, complexity results from the significant influence of soil-structure interaction and the large torsional response, resulting from the varying effective pile lengths, from the landward to the seaward side of the structure. In addition, the interaction of adjacent wharf segments, separated by "movement joints" with shear keys, further complicates the structural response.

317

11.2 MOT DESCRIPTIONS

In general, a wharf is considered to be a structure oriented parallel to the shoreline, while a pier or jetty is perpendicular to the shoreline. A MOT may also include several separate structures such as mooring or breasting dolphins, loading platforms, and approachways or access trestles. Wharves and piers for MOTs may be constructed of reinforced concrete, prestressed concrete, steel, or timber. Composite materials are not addressed in this chapter but are sometimes used as sacrificial piles for vessel impact.

11.3 HISTORICAL PERFORMANCE

Historically, the most widespread cause of seismically induced damage to port and harbor structures has been the liquefaction of the loose, saturated, sandy soils that predominate in coastal areas. Embankment deformations may result in excessive displacement in piles below the mudline. Most "failures" are in fact excessive deformations that disrupt operations and result in damage that is not economical to repair, rather than complete structural collapses that become life-safety hazards, as have been seen in building structures. Werner (1998) documents the historical performance of port and harbor structures subjected to earthquakes in detail.

One specific type of damage frequently observed in earthquakes is to battered piles. Historically, battered piles have been used to resist lateral loads due to mooring, berthing, and crane operations. However, battered piles tend to stiffen the pier or wharf system laterally by a significant amount and result in stress concentrations and shear failure of piles and connections. Structural design of battered piles requires special detailing to account for displacement demand and to provide adequate ductility. In areas of high seismicity, designing pile-supported wharves using only vertical piles is becoming more common.

11.4 STATE OF PRACTICE

Because wharf "failures" are typically the result of excessive deformations, not catastrophic collapse, the state-of-the-practice analysis and design methodologies are based on displacement-based methods rather than the conventional force-based design methods as described in ASCE 7. Structures are typically designed and analyzed to achieve a specific level of performance considering a minimum of two levels of earthquake load criteria.

Design of these structures addresses the complexity resulting from the significant influence of soil-structure interaction and a large torsional response, caused by varying effective pile lengths from the landward to the seaward side of the structure. In addition, the interaction of adjacent wharf segments, separated by "movement joints" with shear keys, further complicates the structural response.

Unique load combinations, such as berthing and mooring, may govern the lateral load design in low seismic regions. UFC 4-152-01 (2017) provides impact velocities for berthing loads. For seismic demand, the dead load plus earth pressure on the structure are considered, with a percentage of the live load added for the maximum dead load case. For the load combinations with mooring and berthing, the earthquake is not considered.

Geotechnical issues are a prime concern for seismic design of these marine structures, with pile foundations often penetrating through weak soil layers. Liquefaction, lateral spreading, and slope stability are all special items that must be incorporated into the analysis and design. Often, these effects cannot be avoided, and the effect of the phenomena must be considered in determining both the structural capacity and demand. In addition, piers and wharves may be close to major earthquake faults, and existing piers may cross faults.

PIANC (2001) contains an excellent treatise on the issues related to the design and construction of wharves and piers in active seismic zones, even though other recommendations in this chapter supersede some of its detailed design procedures and recommendations.

Significant efforts have been undertaken by groups such as the California State Lands Commission, which developed MOTEMS and the Port of Los Angeles (POLA) and the Port of Long Beach (POLB), which funded significant research and development efforts into their own respective seismic design and wharf design codes that are used for the seismic design of marine structures at POLA and POLB. The ASCE 61 Standards Committee continues to work on updates to that document.

11.5 OVERALL APPROACH

State-of-practice analysis and design methodologies are based on displacements and ultimate limit state criteria. A multilevel earthquake approach is used, with varying performance and repairability criteria applicable for each level of earthquake. The two levels of earthquake are commonly referred to as Level I and II, or Operating Level Earthquake (OLE) and Contingency Level Earthquake (CLE).

11.5.1 OLE Performance Criteria

OLE forces and deformations, including permanent embankment deformations, should not result in significant structural damage. The damage would result in only temporary or no interruptions in operations. For new structures, the damage should be visually observable and accessible for repairs.

The OLE return period typically defines an earthquake that is likely to occur during the lifetime of the structure. MOTEMS uses a 50% probability of exceedance in 50 years, or a 72-year return period event.

11.5.2 CLE Performance Criteria

CLE forces and deformations, including permanent embankment deformations, must not result in collapse of the wharf. Controlled inelastic structural behavior with repairable damage may occur. There may also be a temporary loss of operations, restorable within months. For new structures, all damage should be visually observable and accessible for repairs. The global performance of the structure should prevent a "major" oil spill.

For MOTs, a "major" oil spill is generally defined as 1,200 barrels of crude/product. The 1,200 barrels is based on the US Coast Guard's definition of Maximum Most Probable Discharge (MMPD) of oil, used for contingency planning per 33 CFR, Parts 154 and 155. The potential sources of the spill include the flowing and stored oil in pipelines on the wharf/pier and trestle. For the flowing oil, the volume to be considered is the product of the flow rate and the emergency shutdown time to close the system.

The CLE return period typically defines a rare event. MOTEMS uses a 10% probability of exceedance in 50 years, or a 475-year return period event.

11.5.3 Limit State Criteria

MOTEMS, ASCE 61, POLA, and POLB all define structural limit states in terms of maximum strain values for the OLE and CLE. These values are specified for the pile head and in-ground locations. For concrete piles, the values are given for concrete and reinforcing dowels.

11.5.4 Seismic Analysis

The objective of the structural seismic analysis is to verify that the displacement capacity of the structure is greater than the demand for both performance levels. The analysis must consider both inertial loading caused by ground shaking and kinematic loading caused by movement of soil against the piles, where applicable. For irregular configurations, a linear modal procedure is recommended for inertial demand and a nonlinear static procedure for capacity determination. For "regular" structures, a nonlinear static procedure is recommended for both inertial demand and capacity determinations. A nonlinear time history procedure could also be used in lieu of a nonlinear static pushover analysis.

Three-dimensional effects, simultaneous seismic loading in two orthogonal directions, and the full nonlinear behavior of the soil must be included in the analysis. The displacement demand of pipelines relative to the structure should be established to verify their elastic behavior.

The large number of wharf piles, complete with nonlinear soil springs for each pile, complicates the modeling. Several other factors complicate the behavior: inelastic soil springs have different stiffnesses in different directions (e.g., upslope versus. downslope); a significant variation of damping coefficient occurs due to the inelastic behavior of the soil; and the strength of the piles varies due to sequential hinging.

The seismic mass of a wharf or pier structure should be calculated, including the effective dead load of the structure and all permanently installed loading arms, pumps, mechanical and electrical equipment, mooring hardware, and other appurtenances. For MOTs, no additional live load is necessary. The additional hydrodynamic mass of the piles is equal to the mass of the displaced water and is typically very small, with a generally negligible influence on the results.

11.6 EXISTING MARINE OIL TERMINALS

Existing MOTs may be reevaluated for several reasons, such as

(a) Major damage due to an earthquake, vessel impact, fire, or explosion that has seriously degraded the condition of the terminal.

(b) Serious long-term degradation underwater.

(c) Planned major reconstruction of a terminal, or a significant change in operations. Examples include the addition of a large mass on the deck, such as a vapor control system; or a change in the structural configuration, such as an irregular structural modification; to an existing symmetric pile structure. Larger vessels may not necessarily constitute a reason for a reassessment, if mooring and berthing issues are resolved by sufficient changes in operations (limiting wind envelope or reduced impact velocities).

(d) Significant operational life extension of geriatric structures (with more than 50 years of service).

MOTEMS allows reduced seismic criteria to be applied for existing moderate or low risk facilities, with the risk level based on the exposed volume of oil during transfer operations, the number of oil transfer operations per year, and the maximum vessel size.

For a moderate risk facility, the OLE is based on the 65% in 50 year earthquake (48-year return period), while the CLE is based on the 15% in 50 year earthquake (308-year return period).

For a low risk facility, the OLE is based on the 75% in 50-year earthquake (36-year return period), while the CLE is based on the 20% in 50-year earthquake (224-year return period).

Structural and geotechnical information required for a seismic evaluation of an existing facility should be obtained from drawings reflecting current as-built conditions, reports, and codes/standards from the period of construction. If drawings are inadequate or unavailable, a baseline inspection may be necessary. A comprehensive underwater and above water inspection may also be required, along with reconstructed baseline information, if structural drawings are unavailable.

When evaluating existing facilities, component capacities are based on current conditions calculated as best estimates, accounting for the mean material strengths, strain hardening, and degradation over time. The capacity of components with little or no ductility, which may lead to brittle failure scenarios, should be calculated based on lower-bound material strengths.

References

ASCE. 2014. *Seismic design of piers and wharves.* ASCE/COPRI 61-14. Reston, VA: ASCE.

ASCE. 2016. *Minimum design loads and associated criteria for buildings and other structures.* ASCE/SEI 7-16. Reston, VA: ASCE.

33 CFR, Part 154. 2010. *Facilities transferring oil or hazardous material in bulk.* Washington, DC: Coast Guard, Dept. of Homeland Security.

33 CFR, Part 155. 2015. *Oil or hazardous material pollution prevention regulations for vessels.* Washington, DC: Coast Guard, Dept. of Homeland Security.

MOTEMS (Marine Oil Terminal Engineering and Maintenance Standards). 2016. *California code of regulations, title 24, part 2, chapter 31F, 2016.* Sacramento, CA: California State Lands Commission.

PIANC (World Association for Waterborne Transport Infrastructure). 2001. *Seismic design guidelines for port structures.* Lisse, Belgium: A. A. Balkema.

UFC (United Facilities Criteria). 2017. *Design: Piers and wharves.* UFC 4-152-01. Washington, DC: US Dept. of Defense.

Werner, S. D. 1998. *Seismic guidelines for ports, technical council on lifeline earthquake engineering.* Monograph No. 12. Reston, VA: ASCE.

CHAPTER 12

International Codes

12.1 INTRODUCTION

Designers and owners frequently face situations where multiple codes might be used for the same application. This is especially prevalent on international projects, where local codes may differ significantly from a company's own standards. This also occurs when international vendors request exceptions to project-specific criteria to use design codes that are more familiar to that particular vendor or design codes that are embedded into design software.

This chapter provides guidance for such situations, where the design engineer may be required to understand and evaluate multiple seismic design criteria or codes. This chapter does not aim to provide step-by-step guidance for the application of various international seismic design codes.

12.2 CODE CONFORMANCE

Seismic design will generally always require conformance to the governing local building code. Exceptions may exist for large projects, where project-specific design criteria are approved by the regulating agency; however, even in those cases, the company is generally required to demonstrate that its proposed criteria provide an equivalent level of safety as those in the existing local codes.

Examples of typical issues that might arise include the following:

- The local seismic codes were developed for local commercial and residential building construction and are not really relevant for a petrochemical or other industrial project.

- The local code seismic design approach is completely different from the company's own project standards.

- The local seismic design code appears to be less conservative than the company's own standards.

- Conformance to a local code may require design modifications that conflict with basic seismic design principles and may actually result in a less safe design.
- Local vendors only know the local codes, while international vendors do not know the local codes at all.

Each project situation is unique, and no uniform way exists to deal with all these issues. However, the situations noted above are not uncommon, and this chapter attempts to outline some of the mistakes that have been made in the past and some of the cautions to observe when confronting these issues.

12.3 MULTIPLE CODE CONFORMANCE

Many projects may require conformance to both local codes and to project-specific criteria. The design engineer and project management must select an overall approach that is acceptable to both company management and local regulating authorities. Several means are outlined below:

(a) *Compare codes and demonstrate that one is more conservative than the other.* This is very difficult to do in a generic way that covers all possible designs. Section 12.4 discusses several of the issues.

(b) *Perform independent design calculations for each code and demonstrate adequacy for both.* While this is the most complete and thorough method, it is also the most difficult to do within cost and schedule for a project.

(c) *Perform design calculations for project codes and selected checks for local codes.* By conforming to company standards, the company is assured that the design meets its own minimum safety requirements consistent with other projects worldwide. The local check may be viewed more as a regulatory requirement. The extent of checking for local code conformance would need to be negotiated with the regulatory agency.

(d) *Perform the initial design with the company's design contractor and then hire a locally registered company to perform the "final" seismic design.* If this approach is taken, the design engineer must be made aware of any changes to the original design to conform to local codes. It should not be taken for granted that strengthening the original design will always be a conservative approach. For example, if the local code requires only stronger anchorage, the design intent of controlling inelastic behavior may be negated, and an undesirable, nonductile failure mode may result.

(e) *Perform design calculations for the local code and selected checks for the project codes.* This will satisfy the regulatory agencies and may satisfy company requirements, depending on the extent of the difference in codes and the extent of designs that must be checked.

12.4 CAUTIONS WHEN PERFORMING CODE COMPARISONS

When attempting to demonstrate that one design code is more conservative than another, the design engineer must always use extreme caution in applying and interpreting both codes, even when the comparison appears to be straightforward.

12.4.1 Don't Be Fooled by the Return Period

One of the most common errors both engineers and nontechnical parties make is to assume that the amount of conservatism directly relates to the specified return period. This is overly simplistic and should never be used as the sole parameter in comparing codes. However, because the return period is a term widely used to describe low-probability events such as earthquakes, even by nontechnical parties, the engineer must be prepared to thoroughly defend any finding that a code with a shorter return period is more conservative than one with a longer return period.

12.4.2 Compare Both Capacity and Demand

Return periods and associated peak ground accelerations are often used as benchmarks to discuss seismic loading or demand. These are familiar and relevant values to many regulators, even if they are not intimately familiar with seismic design. However, the treatment of spectral shape, adjustments for soil conditions, and other factors are equally or more important.

In comparing the resultant seismic loads from two different codes, the designer must consider the entire demand formulation, including ductility factors, and any other factors that are used in the demand equations. When comparing the overall demand, the basis of comparison should generally be the seismic load coefficient. This is the factor that, when multiplied with the weight of the structure or equipment, will result in the total lateral force or base shear.

Structural capacity calculations also have some variation depending on codes. Additional factors of safety may be applicable, depending on the type of material, the thickness of the section, and the stress mode.

Of primary importance is the need to understand whether the loads and capacities are associated with working stress design or ultimate strength design. In some countries, the loads and capacities are calculated using different code documents.

12.4.3 Know What "Important" Means

Different codes treat importance factors are treated in very different ways. For example, ASCE 7-16 uses a seismic importance factor directly, usually 1.25 for petrochemical and other industrial facilities or for selected items within petrochemical and other industrial facilities.

For comparison, the Venezuelan code, JA-221 (1991), provides a variable return period ranging from 500 to 10,000 years, depending on the degree of risk. The risk classification is a function of the number of people exposed, the potential economic losses, and the environmental impact.

The Russian code, SNiP II-7-81 (1981), uses a K_1 factor that can vary by an order of magnitude depending on the amount of damage that is allowed for the structure or item being designed. In addition, the Russian code defines three categories of facility importance that determine the overall return period to use for design, ranging from 500 to 5,000 years.

12.4.4 Be Aware of Why

Anyone who has been involved in writing or updating model code provisions realizes that building codes do not have purely technical bases. Various provisions result from a combination of technical and political reasons.

For example, code developers may attempt to discourage use of certain structural systems by incorporating factors that penalize the use of these systems with higher load requirements. This is especially prevalent for structural systems generally considered to have low ductility when used in building applications in high seismic zones. Use of the low-ductility systems may be allowed only with a lower R factor, height limits, or other similar restrictions. The designer should be aware of which codes contain these restrictions when comparing the structural requirements.

Engineers are always encouraged to read the Commentary associated with model code provisions to better understand the basis and background for specific provisions. The most recent detailed commentaries are associated with ASCE 7 and the 2015 National Earthquake Hazards Reduction Program Provisions (FEMA 1050). Excellent historical commentary is also available through various editions of the Structural Engineers Association of California's "Blue Book" (SEAOC 1999).

12.4.5 It's Not Just the Capacity and Demand That Matter

Historically, U.S. model building codes were the first to introduce detailing requirements, while other international codes are either silent on the subject or incorporate detailing provisions in later cycles, with a time lag.

The detailing requirements are an integral part of ASCE 7, and extreme caution should be used when attempting to "mix and match" partial code provisions from multiple codes.

12.4.6 Understand the Overall Philosophy

The designer must always understand the context of the international code requirements. For example, long return periods and conservative methods may have historically been used because of low seismicity in previously developed areas of the country. If seismic design has never governed in the past, conservative methods and simple analysis and design methods may have been standard practice but may not be appropriate for projects in new regions without prior seismic design practices.

12.5 VENDOR ISSUES

One of the most difficult issues facing engineers on international projects is dealing with equipment vendors, especially with regards to compliance with project seismic design specifications. Many vendors who do not understand seismic requirements choose to simply ignore them. Others will have specific codes already incorporated into their design software and will request exceptions to the project provisions. Often, the design engineer will learn of these discrepancies when it is already too late to make design changes without affecting project schedules. In those cases, the project engineer must determine acceptability of equipment packages that were not designed to project specifications.

While the vendor is required to provide the equipment package design, the project design engineer is often responsible for attachment or anchorage of the supplied item. In reality, most well-anchored equipment performs well in earthquakes. The walkdown guidance in Chapter 6 can be used to evaluate new equipment packages also. The engineer should focus on those items that might have questionable seismic design characteristics, such as rotating equipment on vibration isolators. The engineer can then work with the project procurement staff to ensure adequate seismic protection, regardless of the seismic design or lack of design by the vendor.

12.6 LANGUAGE ISSUES

Although it may appear obvious, English-speaking design engineers using English translations of international codes often forget that they are not working with the official code and that the English translation is not a legal document.

Design engineers should always have the legal code requirements in their native language at hand and should be encouraged to request clarification from native-speaking engineers who understand the translational nuances that might cause misinterpretation of the intent of some provisions.

References

ASCE. 2016. *Minimum design loads and associated criteria for buildings and other structures.* ASCE/SEI 7-16. Reston, VA: ASCE.

FEMA. 2015. *National earthquake hazards reduction program recommended provisions for seismic regulations for new buildings and other structures—Part 1 provisions and Part 2 commentary.* FEMA 1050. Washington, DC: Building Seismic Safety Council.

JA-221. 1991. *Seismic design of industrial installations.* Caracas, Venezuela: PDVSA.

SEAOC (Structural Engineers Association of California). 1999. *Recommended lateral force requirements and commentary.* Sacramento, CA: Seismology Committee of the Structural Engineers Association of California.

SNiP II-7-81. 1981. *Construction in seismic areas.* Moscow: Russian Industry Standards.

Nomenclature

a = constant

a_p = component amplification factor

A = area of rupture

A_f = pipe anchor and guide forces that are present during normal operation

b = constant

C_d = deflection amplification factor

C_{eq} = limit impulsive acceleration

C_s = seismic response coefficient

D = average fault displacement

D = dead load

D = tank diameter

D_e = empty dead load

D_o = operating dead load

D_s = structure dead load

E = earthquake load (Chapter 5)

E = energy

E = seismic load (Chapter 4)

E = Young's modulus

E_e = earthquake load considering the unfactored empty dead load

E_h = horizontal seismic effect

E_{mh} = horizontal seismic forces including the structural overstrength

E_o = earthquake load considering the unfactored operating dead load

E_{to} = earthquake load for tanks per API 650, Appendix E (2014)

E_v = vertical seismic effect

f_1 = reduction factor for live loads in load combinations

f_2 = reduction factor for snow loads in load combinations

f'_c = specified compressive strength of concrete

f_i = seismic lateral force at level i

f_{yh} = specified yield strength of spiral reinforcement

F_a = site coefficient from ASCE 7-16 Table 11.4-1

F_f = friction force on the sliding heat exchanger or horizontal vessel pier

F_p = component seismic force

F_v = site coefficient from ASCE 7-16 Table 11.4-2

F_y = yield stress

g = acceleration of gravity

G = specific gravity of contained liquid

H = tank liquid height
I_e = seismic importance factor
I_p = component importance factor
L = live load
p = probability of exceedance in t years
P_e = external pressure
P_i = design internal pressure
P_t = test internal pressure
Q = ductility-based reduction factor
Q_E = effect of horizontal seismic forces
Q_{Ee} = horizontal component of E_e
Q_{Eo} = horizontal component of E_o
R = response modification coefficient
R = tank radius
R_p = component response modification coefficient
m_1 = liquid impulsive mass
m_b = body wave magnitude
m_t = total liquid mass
M = magnitude
M_L = Richter magnitude or the local magnitude
M_O = seismic moment
M_{OT} = overturning moment
M_{RES} = resisting moment
M_S = surface wave magnitude
M_W = moment magnitude
n = coefficient = $0.1 + 0.2(H/R) \leq 0.25$
N = number of events of magnitude M or greater
S = foundation deformability coefficient
S = snow load
S_1 = 1.0 s period mapped MCE_R spectral response acceleration
S_a = spectral acceleration
S_d = spectral displacement
S_{D1} = 1.0 s period design spectral acceleration parameter
S_{DS} = short-period design spectral acceleration parameter
S_{M1} = 1.0 s period MCE_R spectral acceleration
S_{MS} = short-period MCE_R spectral acceleration
S_v = spectral velocity
S_S = short-period mapped MCE_R spectral response acceleration
t = time
t_p = bottom plate thickness
t_s = shell thickness
T = natural period of the fundamental mode of vibration
T = return period
T_L = long-period transition period
T_o = thermal force due to thermal expansion

$T_o = 0.2T_S$

$T_S = S_{D1} / S_{DS}$

V = base shear

V_{fdn} = base shear due to foundation

V_{struct} = base shear of structure

V_t = modal base shear

V_{total} = total base shear

w_i = gravity load at level i

W = total seismic weight

W_1 = impulsive liquid weight

W_{equip} = weight of equipment

W_{fdn} = weight of foundation

W_T = total liquid weight

λ = annual probability of exceedance

δ = elastic static displacement at level i due to the forces f_i

μ = modulus of rigidity of rock

ρ = redundancy factor

ρ_s = volumetric ratio of spiral reinforcement

ρ_w = mass density of water

Ω_O = overstrength factor

Index